WHAT SKY TELLS

BY
GÜLDEN BULUT

Translated by Filiz Çiçek

ISBN: 978-1-949872-72-9

Publisher Cosmo Publishing
UNITED STATES OF AMERICA
First Published in 2022 in Turkish

June 2022

GÜLDEN BULUT

She was born in 1977 in İzmir. She graduated from Ege University, the Department of Chemical Engineering. She completed her post-graduate education on "Human Studies In Social Sciences" in Ege University, the Department of Psychology. She continues her education in İstanbul University, in the Department of Philosophy. She is the author of the books *"Mitolojik Astroloji ve Psikoloji, "Külkedisi Pinokyo'yu Seviyor, "Gökyüzünün Söyledikleri"*, and *"Aşkın Simyası"*. Her books *"Külkedisi Pinokyo'yu Seviyor"* and *"Mitolojik Astroloji ve Psikoloji"* are being published overseas under the title *"Astrological Archetypes"* and *"Mythological Astrology ve Psychology"*

CONTENTS

FOREWORD	1
I. THE HISTORY OF ASTRONOMY	5
The Pre-Historic Era	6
Astronomy In Mesopotamia	12
Other Civilizations	17
Islamic Astronomy	21
Modern Astronomy	22
Models Of The Universe	25
II. WHERE ARE WE IN THE UNIVERSE?	29
III. THE SOLAR SYSTEM	33
The Sun	36
The Moon	39
Mercury	41
Venus	43
Mars	45
Jupiter	46
Saturn	48
Uranus	50
Neptune	52
Pluto	53
IV. THE ZODIAC	57
The Tropical Zodiac	58
The Seasons	59
The Sidereal Zodiac	63
V. COORDINATE SYSTEMS	65
Geographical Coordinate System	65
The Ecliptic Coordinate System	68
The Equatorial Coordinate System	69
The Horizontal Coordinate System	71
VI. HOUSE SYSTEMS	73
The Ecliptic Systems	75

The Space Systems 77
The Time Systems 79
VII. PLANETS IN RETROGRADE 83
VIII. PHASES OF THE MOON 89
Eclipses 91
What Are Lunar Nodes? 92
Solar Eclipses 93
Lunar Eclipses 96
Saros Cycles 97
IX. THE STARS 99
The Classification Of Stars 99
The Brightness Of Stars 100
Naming The Stars 101
Star Or Planet? 101
X. CELESTIAL OBSERVATION 103
Measuring Degrees By Hand 104
How Do We Determine The Directions? 105
Observing The Sun 107
Observing The Moon 108
Observing Mercury 109
Observing Venus 110
Observing Mars 111
Observing Jupiter 112
Observing Saturn 112
Observing Uranus, Neptune, and Pluto 113
XI. THE CONSTELLATIONS 115
XII. OBSERVATION TOOLS 125
XIII. TIME AND THE CALENDAR 133
XIV. CASTING A NATAL CHART 141
GLOSSARY OF ASTRONOMICAL TERMS 163
References 171

FOREWORD

> *"...constancy lives in realms above*
> *And life is thorny, and youth is vain"*
>
> *-Samuel Taylor Coleridge, Christabel, 1800*

When the Sun disappears, the sky presents its inimitably beautiful stars for mankind's viewing pleasure. Since we rarely ever look up on dark nights, the sky has long ceased to intrigue us with its lights. It is difficult to see a few bright stars, let alone the Milky Way, in a clear sky with no city lights or air pollution. However, those stars that we now neglect had quite a voice in the formation of the history of mankind.

Prehistoric humans awakened beneath the stars in a world full of mysteries; they watched the Sun, Moon, and other celestial bodies move from one end of the sky to the other throughout the night. People of the antique era, who observed the relationship between the seasonal cycle of the Earth and the motions of the stars, thought that celestial occurrences were intended for us. They laid the groundwork of astrology and astronomy based on a fundamental hermetic tenet: "As above, so below." At the time, astrology and astronomy were two intertwined sciences. Astrologists who studied the planets and stars in Babylonian temples called "ziggurats" were also astronomers.

Celestial observation used to be immensely important for the astrologists of yore. The Babylonians attached particular importance to the emergence of the new moon phase, which signified the beginning of a new month. An astrologist who overlooked it due to neglect or poor visibility faced the risk of being scorned by the king, and this mistake could cost him his post, if not his life. Because the duty of astrologists was not only to interpret the sky but also to observe it. Nowadays, it is no longer necessary to keep such a close watch thanks to ephemerides (celestial journals) and astrology programs; observing the sky has become an obsolete endeavor. We stopped looking up and crammed our natal chart, which is our compass, onto a single A4 sheet. Thus, we lost touch with the cosmos and ended up, in a sense, disconnecting from the spirit of the matter.

Astrologists of the antique era had many specialties alongside astronomy. Because back then, receiving a good education required one to go through a versatile education that included all scientific and intellectual fields; there was no specialization in the modern sense. Al-Battani, one of the greatest astronomers and astrologists of the Islamic world, was also a mathematician. Copernicus, a world-renowned scholar and one of the founders of the Sun-centered model of the universe, was a mathematician, astronomer, and astrologist. Al-Biruni was not only an astrologist but also a historian, a pharmacologist, and a translator. Thales, a pioneer in philosophy and one of the first philosophers in history, was an astrologist. These are only a few examples. Astrologists of yore equipped themselves with astronomical, philosophical, geographical, and historical knowledge. In the antique world, everything related to nature and human life could be made sense of within a whole by using astrology. Over time, however, this whole fell to pieces. Astronomy, philosophy, mathematics, medicine, and other sciences are now evaluated under their own respective titles. Astrology was kicked out of the game like an unwanted player! What is often forgotten is that astrology, like other sciences, emerged and advanced based on substantive observations and rational thought. It also contributed to the emergence of other sciences. Some modern astrologists, on the other hand, ignore these and build their comments solely on astrological keywords, failing to appreciate the profundity of astrology.

Now is the time for aspiring astrologers to learn about other sciences and broaden their horizons! Every astrological comment made without astronomical, philosophical, psychological, geographical, and historical knowledge is incomplete and rudimentary. This book was written for those on track to becoming wellequipped astrologists, with the purpose of meeting this deficit to some extent.

At this point, having a basic understanding of astronomy and mathematics is essential for an astrologist. Because mathematics is the beginning of the road that leads to the universe. Plato, who discovered the harmony of mathematics, had an engraving on the door of the academy he founded that said, "Let no one ignorant of mathematics enter." In this book, I will try to convey astronomical knowledge to you in a straightforward manner, without pulling you into convoluted mathematical equations. I also saw fit to include some astronomical knowledge that falls beyond the scope of the link between astrology and astronomy. Because comprehending any celestial body or event is actually a precursory step toward understanding astrological terms. Knowing that the galactic center points to the constellation Sagittarius and that Jupiter moves in its orbit as the Earth's insurance will not only help astrology students understand Jupiter and Sagittarius and so internalize astrology, but will also present them with a new perspective.

In addition, I will cover a variety of topics in this book, ranging from the history of astronomy to celestial observation, from analyzing astrological data in an astronomical light to casting a natal chart. As we study the sky, we will see how little we know about it, but we will also be getting closer to the common legacy of mankind.

People of the antique era had to listen to the sky and the earth amidst their fight for existence; they observed the Sun's motions, determined the time, and created civilizations. Throughout history, humans have been influenced by the sky, and in turn, the sky has been influenced by humans. Stories of the earth are, in a sense, stories of the sky. The unity of the earth and the sky is a unity that shapes and steers our lives.

A sky full of stars on a clear night is intriguing not only for an astrologist who makes sense of it, but for every individual. Let's get to know the sky, which has guided mankind for millions of years, and look up. Let's try to read the skies with informed eyes from our galaxy, which is but a speck in the vastness of the universe.

Gülden Bulut
June 30, 2016, İzmir

I.
THE HISTORY OF ASTRONOMY

"The One remains, the many change and pass;
Heaven's light forever shines, Earth's shadows fly."

-Percy Bysshe Shelley, *Adonais*

The origins of astronomy date back to prehistoric times. Almost all cultures have taken an interest in the objects and events in the sky. Stars that appeared to be falling toward the Earth (i.e., meteorites), the motions of the Sun and Moon, the halos that formed around them, bright balls of fire (i.e., comets), and many other phenomena were all sources of mystery for the people of the era. Even though the sky they'd be looking at isn't the same as ours, it still creates the same magnificent feeling. The only difference is our thoughts about the sky!

People of the antique era realized that there was an order in the sky, just as there was on Earth. They used the calendar in the sky in order to know when migratory wild animal herds would migrate, to predict the arrival of rain and cold weather, to prevent annual floods, to carry out agricultural activities (such as knowing when wild grains would mature), and to determine religious days. They built the foundations of astronomy by not only detecting the Sun's location but also directly observing the effects of the seasons on their own lives and recording these observations. They kept records of celestial events by writing about and making drawings of astronomical events on cave walls, papyruses, and clay tablets. Based on these records, they created calendars. They made sense of the sky, which was both fascinating and terrifying in its vastness, by approaching it mathematically. They ensured the advancement of astrology and, hence, astronomy.

Now, let's have a look at how astronomy has developed throughout history. We'll come across astrology along the way, because they're two fellow travelers who set out together and parted ways later!

The Pre-Historic Era

Our Earth is at least four and a half billion years old. Our story began around 200,000 years ago with Homo sapiens, which means "wise man." Just like us, prehistoric humans also needed food, shelter, and protection. They lived in caves in small groups because they didn't know how to build houses yet. They conveyed to us their unwritten history by leaving marks and traces where they dwelled.

In the "Hunting and Gathering" era (about 30,000 or 40,000 years before our time), our ancestors could use sharp tools and make drawings and sculptures. Our understanding of the nature of humanity's relationship with the skies dates back to around this time. Archaeological and anthropological discoveries about primitive tribes living around 30,000 BC show that cyclical events in the skies were documented.

One of these findings is the Ishango bone, discovered in Ishango, Central Africa, and thought to date between 20,000 and 18,000 BC. There are three rows of tally marks on the bone. Another similar example is the Blanchard Bone Plaque, a bone plaque discovered in Southwestern France that is engraved with tally marks and animal figures. To this day, the meaning of the figures engraved on the bone is a much debated subject. From crescent to full moon, the Moon's everchanging phases must have piqued people's interest, but why did our ancestors record celestial events?

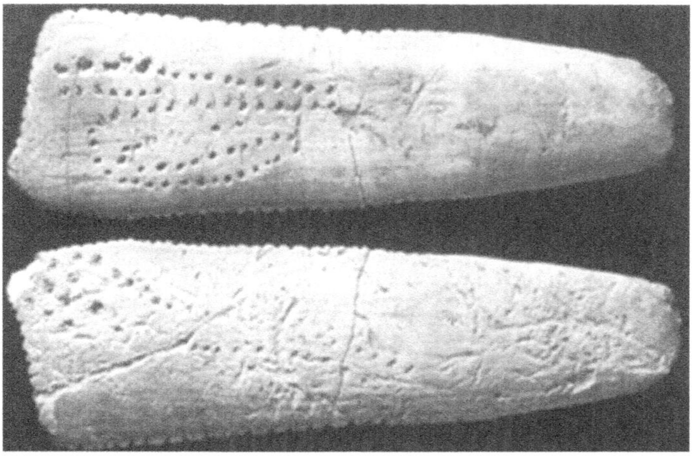

Image 1. The Blanchard Bone, Courtesy Harvard University, Peabody Museum

And why did they take an interest in the skies to begin with? Archaeoastronomer Alexander Marshack claims that these tally marks are a part of some sort of lunar calendar. He adds that this calendar allowed people to predict when animals would migrate and when plants would bear fruit.

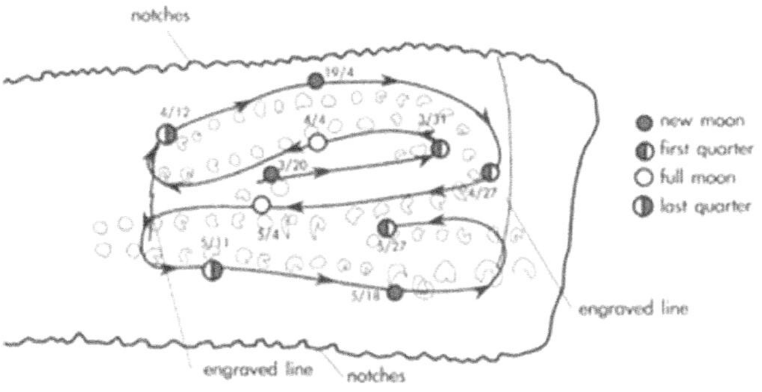

Figure 1. Depiction of the mark carved on the Blanchard bones

Image 2. The Venus Of Laussel, The Neandertal Museum

Another significant remnant is The Venus of Laussel, a bas-relief discovered at the entrance to a rock shelter in Dordogne, France. Venus is depicted holding a horn in her right hand in this bas-relief from around 25,000 BC. There are 13 lines on the horn. The lines in question are thought to represent the 13 months in a solar year or the monthly rhythm of the ovulation period and the cycle within women's bodies.

There are also comments suggesting that the drawings discovered on the walls of the Lascaux cave in Dordogne, France, in 1940, are related to the skies. These drawings are thought to be from the same era as the aforementioned bas-relief. The geometrical shapes on the cave's walls are thought to depict the motions of the Moon and stars. Comments about how these figures depict the three bright stars known as the "Summer Triangle[1]" and the Pleiades star cluster reveal more about the influence of the skies in the lives of the people of that era.

Humanity has also left significant marks dating back to the Mesolithic era, a period of transition. Much like the people of earlier eras, the hunter-

1 When we look up at the sky during the summer months, we can see three bright stars: Vega, Deneb, and Altair. When we connect these three stars, we get a triangle shape. This shape is called the "Summer Triangle."

gatherers of this era continued to dwell in caves. However, the animals that they hunted had migrated because of climate change, leading to the survival of less agile animals with thinner skins. This is why people of the era lived in groups.

Image 3. Göbekli Tepe, Şanlıurfa

We see large monuments built by the people of this era that raise astronomical and astrological questions. One of these monuments is among the most significant discoveries that have shed light on the history of humanity: "Göbekli Tepe," discovered in Urfa in 1995 by German archaeologist Klaus Schmidt. A sacred space spread out over 90,000 square meters that dates back to 9500 BC—11,500 years before our time—and has been used for over 2000 years! Much remains an enigma about these structures built by humans who still dwelled in caves, lived nomadic lives, and hadn't yet transitioned into a settled life.

One of the main features seen in temples is two T-shaped obelisks at the center, surrounded by 10 to 12 obelisks.[2] In one of these center obelisks, the Moon is depicted as a crescent, and the Sun is depicted with a hollowed-out center. The issue arises as to whether Göbekli Tepe was built as an observatory to watch celestial bodies and events. Could these structures have anything to do with astronomy or even astrology? People

2 T-shaped obelisks can be found not only in Göbeklitepe but also in other Neolithic centres near Göbeklitepe, such as Nevali Çori, Hamzan Tepe, Sefer Tepe, and Karahan Tepe.

of the antique era may have understood that the Sun doesn't follow the same path every day, that the durations of the days vary, and that there is a relationship between the motions of the Sun and the seasons. Furthermore, some researchers suggest that the obelisks may have been used to determine the equinoxes and solstices. There are also some comments stating that the animal figures on the columns may represent the constellations. For instance, it is believed that a scorpion figure on one of the columns in Göbekli Tepe indicates the location of the constellation Scorpio on the zodiac belt. The angle between the head and the wings of the eagle and the vulture around the scorpion are in cohesion with the angle between these constellations.

Now consider this from a different angle: this sacred area was used for rituals. This discovery shows that it was not the taming of animals and plants that led people to live settled lives; it was their religious beliefs. Based on this, we can conclude that prior to the transition to settled life, people were organized around a leader in religious centers.

Furthermore, there are motifs on some of the obelisks in Göbekli Tepe that depict the union of man and woman, which brings to mind the sacred marriage ceremonies that took place in Sumerian temples. This raises the possibility that the mythology of the era may have merged with the people's knowledge of the skies. This shows that religion, mythology, astronomy, and possibly astrology have advanced on common ground!

Finally, let’s talk about Stonehenge (3000 BC), which is another famous example of the large stone structures known as Megaliths. But first, let's consider the characteristics of the relevant period. Humans abandoned caves and the nomadic lifestyle during the Neolithic period, instead settling in areas that were near water and convenient for crop cultivation. They opted for a continuous life; they tamed animals and plants and began to engage in agricultural activities around the places where they dwelled. Agriculture required them to know about the times of the seasons beforehand—i.e., it required them to have calendar knowledge. A calendar meant knowing and understanding the motions of the celestial bodies. Astronomy began to develop along with agricultural activities. There was, in fact, a reciprocal cycle. Understanding the cycle of the skies helped propel agriculture forward.

Stonehenge, made up of stone monuments erected in a circular fashion, is believed to have been built for the purpose of observing the skies. This theory is supported by strong data: analyses of the meticulous measurements taken on the stone circle point to the conclusion that this stone structure was actually an astronomical observatory. In fact, these analyses suggest that this contraception could be used not only to establish the rising and setting times of the Sun and Moon—i.e., creating a seasonal calendar—but also as an observatory for predicting upcoming solar and lunar eclipses. So, based on centuries of observation, the people of the Neolithic era may have realized that eclipses only occur when the Moon and the Sun are in specific positions!

Image 4. Stonehenge

The working principle of the contraption at Stonehenge is fairly simple. The Sun doesn't rise or set from the same spot every day. You can also easily photograph this.

Pick a few reference points to anchor the photo. Make sure you choose a location where you can see both the east and the west. On March 21, aim your camera due east. Press the shutter during sunrise and sunset every 10 days. Then overlay all the photos using Photoshop. The rising point of the Sun will leave a trace. This was, in fact, what the people of the antique era were doing; the only difference was that they did this with stones. They determined time based on the shadow cast by the Sun amid the stones, which was, in a sense, similar to a sundial. For instance, they observed that on March 21, the Sun rose from between two stones. They were, in fact, using a horizon calendar[3] to determine the time.

Astronomy In Mesopotamia

Astronomy and astrology spread throughout the world from Mesopotamia. Among Mesopotamian civilizations, the Sumerians, Babylonians, and Chaldeans have played a major part in the development of astrology and astronomy.

The Sumerians

The Sumerians, one of the world's earliest known civilizations (4000–3500 BC), emerged in Mesopotamia, between the Euphrates and Tigris rivers. The zone was divided into two: Upper Mesopotamia and Lower Mesopotamia. Compared to Upper Mesopotamia, Lower Mesopotamia had more alluvial and fertile soil. Rivers flooded frequently in Lower Mesopotamia, leaving the soil awash. The farmers got better yields from their fields when the waters ebbed. However, the rivers did not flood on a regular basis, and these sporadic floods caused significant damage to agriculture in general. The Sumerians' research on how to control irregular flood waters led them to use mathematics and geometry.[4]

3 Here is what a horizon calendar tells us: throughout the year, the Sun rises from various points on the horizon that are within approximately 45° of due east. On March 21 and September 23, the Sun rises due east and sets due west. For an observer in the northern hemisphere, the Sun rises in the northeast on June 21 and in the southeast on December 21.

4 We know from cuneiforms that the Sumerians had advanced mathematical knowledge in their era, that they were able to solve rather complicated arithmetical and geometrical problems, that they knew algebra to a certain extent, and that they used the decimal and the sexagesimal number systems.

The Sumerians attributed godlike qualities to the planets they discovered, and they saw the planets' motionsas a result of the events on the ground. With these analogies, they named the archetypes, which are humanity's common heritage, in celestial bodies, and divine mythologies shaped their lives.[5]

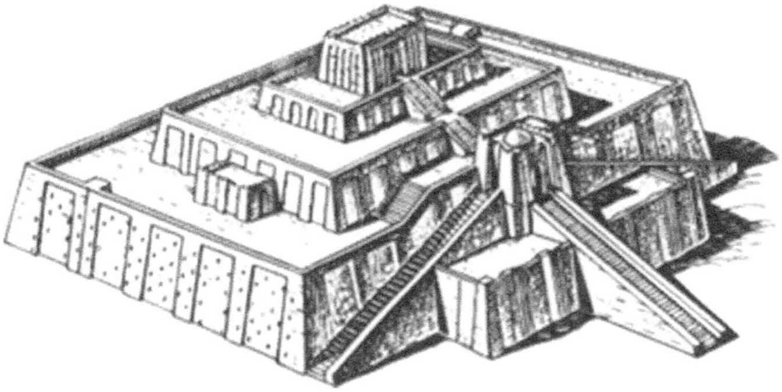

Image 5. Depiction of a ziggurat

The Babylonians

Although we do not have direct access to Sumerian astrology and astronomy sources, we do know from written Babylonian documents that the Babylonians were the first civilization to closely observe celestial bodies.

The Babylonians came to Mesopotamia in 2000 BC. After nearly a century of living under Sumerian and Akkadian rule, they founded the city of Babylon.[6] The Babylonians, like the Sumerians and Akkadians, observed the sky from ziggurats. Known as "stepped temples," the

5 Those interested in learning more about the connection between the mythologies of past civilizations and astrology can get my book, "Mythologic Astrology and Psychology," published by Zodyak Astroloji Publishing.

6 The Babylonians were Amorites, who were a Semitic tribe. The reason we know them as "the Babylonians" is that they founded and developed their civilization in Babylon, and the capital of the empire they built was the city of Babylon.

ziggurats were built as concentric squares rising upward. The ziggurats[7], where the foundations of astronomy and astrology were laid, functioned not only as archives and libraries and places to pray, but they were also used for watching the stars. Other celestial events, such as solar and lunar eclipses, were also recorded in ziggurats.

The Babylonians were the first to discover and name five of the planets, excluding the Moon and Sun. They also created the concept of the week by naming the days after the Sun, Moon, and other planets they observed. They discovered Mercury, Venus, Mars, Jupiter, and Saturn, and they named these planets after their own gods.

The Babylonians were able to calculate when and where the planets would emerge and disappear. They were particularly interested in Jupiter's motions, but they also observed the motions of all celestial bodies and documented their findings on clay tablets. "Enuma Anu Enlil"[8] (1700 BC) is a tablet describing the motions of the Moon and Sun, weather, and the relationships between the planets and stars. Furthermore, the tablet explains how to calculate the day the Moon will first appear and where to observe lunar and solar eclipses.

The Babylonians discovered a connection between the Sun and the phases of the Moon; they realized the Moon reflected the Sun's light. They also discovered that the stars seen at various times of the year differed from one another.

7 Again, in this era, the use of temples as observatories emphasizes the function of religion. Priests were not only clergymen but were also involved in the structuring of political, social, and economic order. Although wars caused kings' powers to be shaken and heroes to emerge, the power of clergymen remained unchanged for a long time. This is one of the reasons that religion, astronomy, and astrology are intertwined.

8 "Anu" is the god of the sky, and "Enlil" is the god of the earth. "Enuma Anu Enlil" is the book of the gods of the sky and the earth.

Image 6. A replica of the Venus tablet of Ammisaduqa, made around 650 BC. The original Venus tablet of Ammisaduqa was made around 1600 BC.

Believing that the stars were fixed onto the celestial sphere, they created the constellations by likening stars that were close to each other to certain shapes in order to find them and keep them in mind easily.[9]

"The Venus Tablet Of Ammisaduqa," also mentioned in Babylonian records, contains prophecies and observations regarding the cycle of Venus.[10]

The Chaldeans

It was the Chaldeans[11] who established astrology in its current form. The most magnificent ziggurat was built in Babylon under the reign of the Chaldean Empire in the second half of the 7th century. The Chaldeans predicted the planets' future motions based on their past motions, and

9 It is said that shepherds came up with imaginary shapes and names for the constellations in order to find the best time to move their herds elsewhere for grazing.

10 The depiction of Venus as a pair of crescents represents, in a sense, the phases of Venus.

11 The city of Babylon regained its power when the Chaldeans ended the Assyrian reign in Babylon and established the Chaldean Empire. This period where Babylon rose back to power with the Chaldeans is called the "Neo-Babylonian period."

they spearheaded the expansion of astrology to encompass individuals by calculating the planets' future positions. Astronomy improved as a result of Mesopotamian astrology and mythology.

The Egyptians

The birthplace of the Egyptian civilization was the Nile Valley. The Nile River flooded regularly. The waters began to flood in July and ebbed in November. The Egyptians learned about the seasons and the best times to plant crops based on regular floods. They integrated this knowledge with their knowledge of the sky and realized the significance of observing the sky.

The sky helped them orient themselves. They learned about ploughing the soil, planting seeds, and harvesting crops by using the calendar. This is the first calendar devised by civilized man that is based on the seasons and the Sun.

The Egyptians also devised the Sothis calendar. Sothis was the last star to lose its brightness. Thanks to Sothis (Sirius), they were able to determine the deviation of the floods of the Nile. During the Nile's flooding season, all the stars above Cairo's latitude dimmed as morning approached. The calendar began around midJuly, which was when the Nile flooded. This shows just how advanced the Egyptians' knowledge of astronomy was. Furthermore, analyses of the pyramids revealed some information indicating that there was certainly some astronomical guidance in the construction of the pyramids.

The Karnak Temple in Egypt, dedicated to the god Amun, is one of the largest temples in the world. The temple was built to face the setting Sun during the summer solstice. The guiding role of the skies in the establishment and development of civilizations is emphasized repeatedly.

Other Civilizations

The other civilizations' fascination with the sky never ceased, either. People in Central Africa used lunar observations to forecast the weather. The indigenous peoples of North America positioned their houses based on where the Sun would rise. They also built their sacred spaces with the four directions in mind. They even shaped their graves according to the daily changes in the skies. The Mayans (3379 BC) positioned their structures to face Venus, and they created their calendar based on their observations of Venus.

Around 5000 years ago, the Chinese began recording lunar and solar eclipses, comets, meteorites, and sunspots. They were the first to collect data on sunspots. Another first occurred in 1054, when the Chinese recorded the Crab Nebula supernova explosion. There is, however, still no scientific explanation for how they obtained this information.

As it is understood, the first civilizations' observations of the sky necessitated a high degree of knowledge in mathematics, architecture, and astronomy. The observation and recording of celestial events contributed to the advancement of astronomy.

Although people have been fascinated by celestial events and motions since the early ages, they initially only observed the sky. They later attempted to deduce the causes of these events as well as the connections between them. The Ancient Greeks were the first civilization to concentrate more on the causes of astronomical events than on the events themselves.

Image 7. A Mayan observatory
(The structure has windows that are suitable for observing Venus)

The Ancient Greeks

The Greek civilization is the product of the Cretan and Mycenaean civilizations, which owe their cultures to the Mesopotamians and the Egyptians. Their presence was first felt on the shores of the Aegean Sea around 1000 BC. Mesopotamian astrology was unable to progress beyond observing the apparent motions of celestial bodies and addressing their correlations. It was the Greek civilization that first sought to understand nature by drawing on the mathematical, astronomical, and astrological knowledge and experience of past civilizations and asking questions about nature: why were the motions of the Sun and the Moon different from those of the other stars? Why and how often did solar and lunar eclipses happen? Following in the footsteps of previous civilizations, the Greeks began to investigate the causes of these events.

Thales was the first philosopher in Ancient Greek history. During the transition from mythological thought to rational thought, Thales proposed the hypothesis that the universe was made up of water, which was the first scientific view in history. This effort to explain the universe, which began with Thales, lasted until 400 BC. Thales used geometry to study the apparent motions of the Moon and Sun. He correctly predicted a solar eclipse that occurred in 585 BC. Anaximander, a student of Thales, devised an astronomical and geographical calendar and calculated the distances of the Sun, Moon, and planets from Earth.

Pythagoras was the first philosopher and mathematician to conduct experiments in his work. Based on the phases of the Moon, he proposed that the Earth is spherical. He also noticed that the curvature of the terminator, which is the line separating the light side and the dark side of the Moon, changed along with the phases of the Moon. He deduced from this that the Moon was spherical rather than flat.

Although Ancient Greece produced one thought system after another, Plato (399 BC), a student of Socrates, holds a significant position. Plato merged his thoughts with Pythagorean traditions within a system and put them in writing. He advised astronomers to look not at the sky, but into their own inner worlds and minds. The mind would show the people that the starry sky revolved around the Earth in a circular motion. Because that was what befitted celestial bodies. He said, *"All of us tell lies about those great gods, the Sun and the Moon... We say that they and diverse other stars do not keep the same path, and we call them planets or wanderers... Each of them moves in the same path—not in many paths, but in one only, which is circular, and the varieties are only apparent."* According to him, this circular motion was perfect. Although his natural philosophy may appear mathematical, it included mythology.

Aristotle, a student of Plato, devised a closed system to explain the entire universe. The universe that he had designed was mechanical, despite being hierarchical and divine. According to Aristotle, the universe was made up of concentric spheres, and the Sun, Moon, and planets rotated in concentric, circular orbits around the Earth. He also believed that there was a power that ruled and surrounded the entire universe: the "unmoved mover." He also claimed that the substances of which the skies were made were of a different quality than the substances of which the Earth was made. He claimed celestial bodies were infinite, unchanging, and incorruptible, whereas the Earth was situated in the sublunary sphere, where things were temporary and flawed.

Aristotle believed that the Earth was made up of four elements, each with its own unique properties: earth was dry and cold, water was moist and cold, air was moist and hot, and fire was dry and hot.

Astronomically, Aristotle was able to explain the phases of the Moon correctly. He confirmed that the Moon spread the light emitted by the Sun. Based on the appearance of the Earth on the Moon during lunar eclipses, he proposed that the Earth was a sphere.[12] Although Aristotle's views on astronomy were incorrect, they were widely accepted until the Renaissance, causing astronomical developments to come to a halt.

The 300-year-long Hellenistic period began when Alexander the Great brought Mesopotamia under his reign around 300 BC. This period also marked the beginning of a process of change in astronomy and astrology. Berossus of Chaldea, a priest, chronicler, and astrologist, was sent to Greece, where he founded a school of astrology on the island of Kos. In this school, the Sumerian, Chaldean, and Babylonian creeds were conveyed to the west. Thus, the Greeks were introduced to the astronomical knowledge of other civilizations. The findings led to the conclusion that the Earth was round. Aristarchus (312–230 BC) was a

12 Celestial bodies were always considered spherical because of religious beliefs. The most perfect object is the Earth, and all the other celestial bodies are similar to it.

philosopher who held views that were far ahead of their time. He was the first to suggest the heliocentric model. According to Aristarchus, the Sun was at the center of the universe, and all planets, including the Earth, revolved in a circular orbit around it. Aristarchus's discovery was forgotten until the Copernican Revolution, which was about 1500 years later.

Hipparchus, who lived in the 2nd century BC, was another important philosopher. He compiled the coordinates of about 850 stars and categorized them based on their brightness. This classification is still in use today. He calculated the distances of the Sun and the Moon with the star catalogues that he prepared based on highly precise measurements. He devised the method of determining the location of any point on Earth by using the circles of latitude and longitude. He detected the irregularities caused by the motions of the Moon and the gravitational pull of the Sun. Hipparchus compared his own records with earlier records and saw that there was a systematic shift in the coordinates of all the stars. He proposed that this systematic shift was caused by the Earth's precession movement. Interestingly enough, this finding is entirely correct. Hipparchus was among the most prolific astronomers of his time.

Another significant astronomer and astrologist who emerged in the final period of the Alexandrian school was Ptolemy. Ptolemy was also regarded as one of the greatest geographers of the ancient world. Among his writings are the 13-volume astronomy encyclopedia "Almagest[13]" and the 4-volume astrology treatise "Tetrabiblos." Ptolemy devised a system of the universe based on Hipparchus's discovery, and this system remained valid for 1400 years. Ptolemy explained the structure of the universe with his

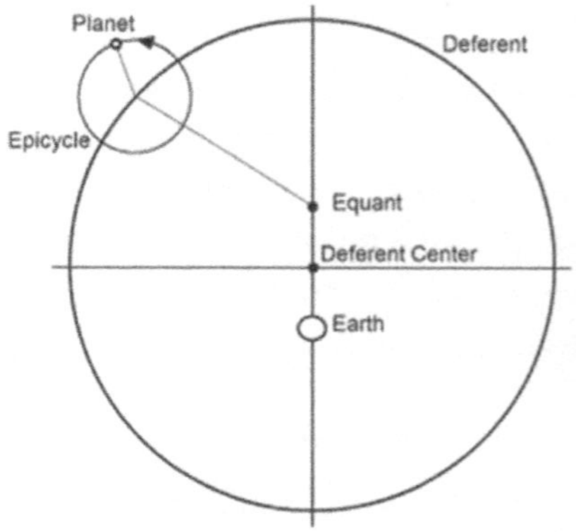

Figure 2. The model by which Ptolemy explains the orbital motions of the planets

13 The Almagest lists 48 constellations. A large portion of the book features mathematical methods regarding the motions of the planets. Ptolemy's views on the motions of the Sun doesn't differ greatly from that of Hipparchus. However, he devised a theory on the motions of the Moon that was superior Hipparchus's theory.

model of the universe, with the Earth at its center and the Moon, Mercury, Venus, the Sun, Mars, Jupiter, and Saturn revolving around it in that order. He was able to explain the irregularities in the motions of the planets by devising the geocentric system and creating complex orbits.

Islamic Astronomy

The 100-year period between the end of the ancient civilizations and the beginning of the Italian Renaissance (between the 4th and 13th centuries AD) was the Dark Age of Europe. In the time of the Romans, who came onto the stage of history after the Greeks, philosophy and science came to a standstill. Around this time, the Arabs integrated their conquests with other civilizations' knowledge, making significant progress in science and philosophy. The nomadic lifestyle of the Islamic world, as well as the fact that they kept records of celestial bodies in order to calculate directions and therefore carry out their religious duties, played a role in this.

Among Islamic astronomers, two names stand out: Muhammad Al-Battani and Ibn Yunus. At-Battani was able to calculate the precession of the vernal point with greater accuracy than Ptolemy, and he devised a new set of astronomical tables. Ibn Yunus gained recognition for his observations of the Sun and the Moon in Cairo.

Islamic astronomers followed in Aristotle's footsteps; they believed that the Earth stood immobile at the center of the universe, and all the other celestial bodies, including the Sun, revolved around it at a fixed speed in circular orbits. They all accepted Ptolemy's view as it was.

Furthermore, Islamic astronomy established the seasons' durations based on the apparent conditions of 28 constellations, as well as the geographical directions to use in their night journeys based on the apparent conditions of the Moon and some visible stars. Many astronomical terms and star names used today are in Arabic.[14]

---

14 Names of some of the brightest stars such as Rigel, Aldebaran, Algenib, Algol, Deneb, and Vega; fundamental astronomic terms such as zenith, nadir, and azimuth; and again, fundamental observation tools such as alidade, astrolabe, and theodolite are in Arabic.

The Islamic world's most important contribution to astronomy was that they founded observatories in the modern sense. In these observatories, they devised new observational tools and trained many astronomers. Taqi ad-Din al-Rasid, one of the best astronomers of the Ottoman era, founded the first observatory in Istanbul. He observed a comet that appeared in the sky on September 11, 1577 and was visible for about a month. This observation was one of the earliest scientific studies of the observatory. Taqi ad-Din prophesied to the monarch of the time that this event was a harbinger of victory against the Persian army. Following the deaths of many people in the plague outbreak of 1578, a fatwa was issued for the observatory to be demolished. The observatory was destroyed by cannon fire on January 21, 1579, putting an end to advances in astronomical studies. However, the research conducted in the observatory founded by the Dutch astronomer Brahe in the same era revealed brand-new horizons in astronomy.

Modern Astronomy

Modern astronomy began with the Renaissance. Beginning in the late 16th century, this was the era of modern science, based on experiments and mathematics. Astronomical breakthroughs are associated with the individuals of this era.

Copernicus (1473-1543): Although Copernicus represents a turning point in the history of thought, in many ways he adhered to Aristotle's views. He did, however, make an important contribution by placing the Earth at the center of the universe and proposing a heliocentric system. He voiced his views as follows: *"We sail out of the harbor, and the countries and cities recede. In that case, the sphere of fixed stars only appears to move; it is not the sphere of stars that makes one rotation each day, but rather the terrestrial sphere itself, which appears immobile to us."* Despite these thoughts, he concluded that the Sun was not exactly at the center, but near it, because he could not fully observe the circular motions of the Earth and the other planets around the Sun. Despite his heliocentric model, he took the model in Ptolemy's *Almagest* as reference.

Copernicus was a clergyman. He thought that the church would turn against him, so he didn't have his book published until the last days of his life. However, one of his disciples, Giordano Bruno, ardently defended Copernicus' views, which led to his execution by burning at the stake in Rome in 1600.

There are two scientists who had more of a voice at the beginning of the modern era that followed the Copernican revolution: Tycho Brahe, a great observer, and Johannes Kepler, a unique astronomer.

Tycho Brahe (1546-1601): Brahe designed tools of measurement through systematic methods and measured the orbits of the planets. According to Brahe's model, the Moon and the Sun revolved around the Earth, and the other planets revolved around the Sun, all moving in circular orbits. Although Brahe's research showed that Ptolemy's model of the universe was incorrect, he never embraced Copernicus's theory due to his commitment to the church. Nonetheless, the tools he designed made significant contributions to astronomy.

Johannes Kepler (1571-1630): Kepler[15], who served as Brahe's assistant in 1600, took on the duty of examining Mars in Brahe's research group. According to him, Mars moved in an elliptical orbit with the Sun at its focus. Kepler generalized this discovery to include all planets, and after 20 years of countless observations and measurements, he established three laws:

1. All planets move around an elliptic orbit with the Sun at one focus.
2. As the planets move in their orbits, they do not sweep out equal areas in equal amounts of time, but a line segment joining the planet to the Sun sweeps out equal areas in equal amounts of time.
3. The cube of a planet's mean distance from the Sun is proportional to the square of its period of revolution.

15 Kepler believed that there was a certain harmony between the universe and humans. He was also known for his astrology-based predictions.

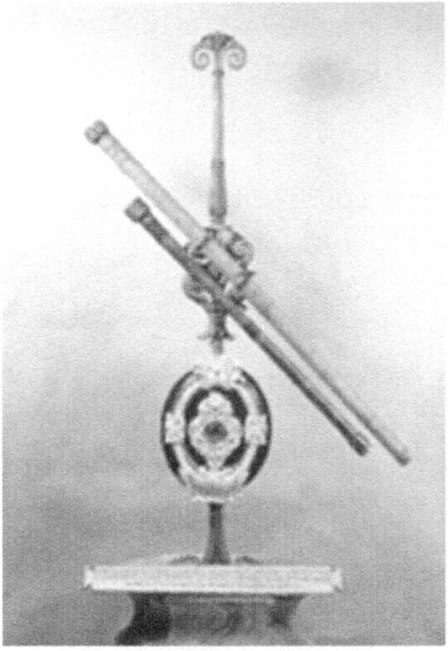

Image 8. Galileo's telescope

Galileo Galilei (1564-1642): Galileo Galilei is another significant touchstone in the history of modern astronomy. Galileo, who learned in 1609 that a Dutch optician had invented a lens that magnified distant objects, made the first telescope by utilizing the phenomena of reflection and refraction of light. When he pointed his telescope at the sky, he was able to see much fainter celestial bodies than he could with the naked eye. He noticed that the Sun's dark spots moved, and that the Moon's surface was not as perfect and smooth as previously thought, but was partially rocky and covered with mountains and valleys. It was, again, Galileo who discovered the four large satellites of Jupiter, as well as the one who detected the phases of Venus and proved that Venus revolved around the Sun. The church condemned him for claiming that the Earth revolved around the Sun. The Inquisition sentenced him to life under house arrest in 1630. He spent his final years blind and alone, under the watchful eye of the church. Galileo's use of the telescope paved the way for astronomical advancements.

Isaac Newton (1643-1727): Newton, who was born in the year of Galileo's death, discovered the causes of Kepler's laws. Newton provided a mathematical explanation for the interaction between celestial bodies using the laws of nature that he found regarding the law of motion and gravitational force. Newton discovered that the gravitational force between two objects is directly proportional to the product[16] of their masses and inversely proportional to the square of their distance. This theory debunked the geocentric model of the universe and marked one of the most important, if not the greatest, step taken in astronomy. After Newton, astronomy advanced further with each passing day.

16 In mathematics, the result of a multiplication is called "product."

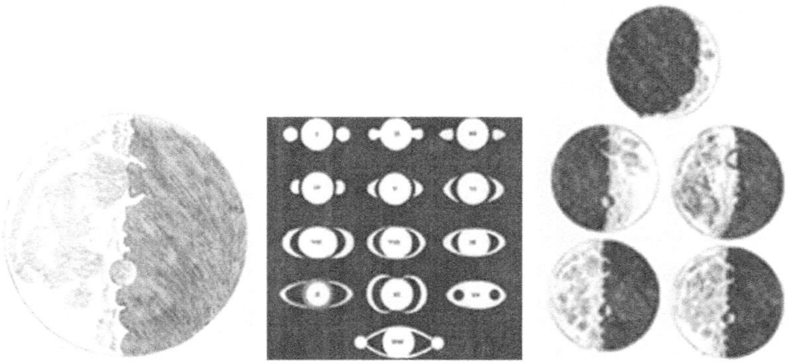

Image 9. Galileo's drawings of the Sun, Saturn and the Moon

The process up to the present day has seen the discovery of trans-Saturnian planets and their satellites, as well as thousands of asteroids and comets. Telescopes and the other techniques were used to determine the distances, temperatures, and chemical compositions of celestial bodies. In 1924, the Hubble telescope revealed the presence of galaxies other than the Milky Way. The first manned Moon landing took place on July 20, 1969. The evolutionary process of the stars was expanded upon. It was discovered that the universe was constantly expanding. Every day brings new discoveries that expand our understanding of the universe, and every day we lift a little more of the endless cloak of secrecy.

MODELS OF THE UNIVERSE

Throughout history, humans have attempted to understand the universe and find their place within it. Every culture has developed its own understanding of the universe. Theoretically, models of the universe can be divided into two classes:

Earth-Centered (Geocentric) Model

Complex explanations were developed based on this viewpoint to understand the daily revolution of the Sun, the planets, and the stars around the Earth. In this sense, Eudoxus of Cnidus devised the first geometrical model of the universe. His model of the universe was made up of concentric spheres, with the Earth at the center. The Sun and all the other planets were attached to spheres surrounding the Earth. As the spheres rotated, so did the planets.

Among these models, Ptolemy's had the most enduring influence. He developed the geocentric model of the universe and explained the complex motions of celestial bodies in his work titled *Almagest.* This theory was accepted without dispute for about 14 centuries.

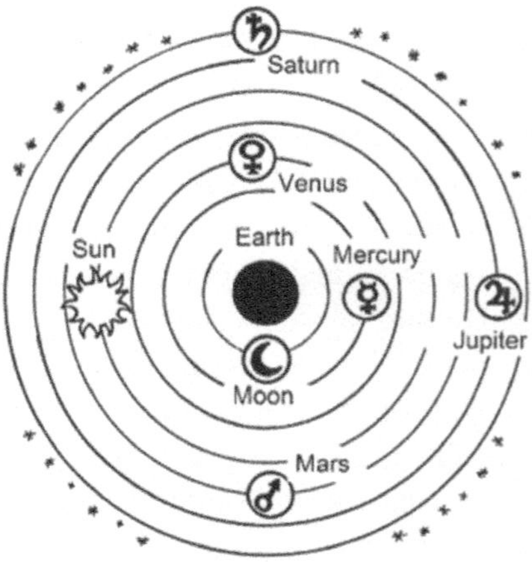

Figure 3. Ptolemy's geocentric model of the universe

Sun-Centered (Heliocentric) Model

Although the heliocentric model of the universe is associated with Copernicus, it was first proposed around 300 BC by the Greek astronomer Aristarchus. His theory, however, was not accepted due to the social circumstances of the time.

Copernicus thought that the existing astronomy tables were inadequate, so he conducted numerous observations from the steeple of the cathedral.

Much later, he proposed a theory that would debunk the church's official view of the universe. He attempted to comprehend the structures of celestial bodies through Ptolemy's model of the universe. In light of these thoughts, he devised a Sun-centered system in which the Earth also revolved around the Sun.

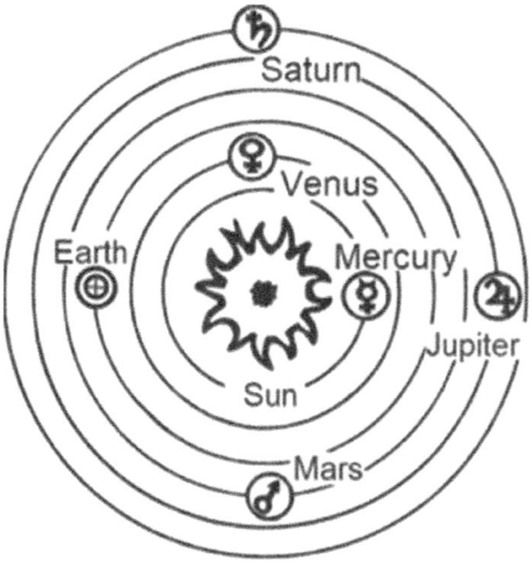

Figure 4. Copernicus's heliocentric model of the universe

II.
WHERE ARE WE IN THE UNIVERSE?

"...Others think that although their number is not without limit, no number can ever be named which will be greater than the number of grains of sand. But I shall try to prove to you that among the numbers which I have named there are those which exceed the number of grains in a heap of sand the size not only of the Earth, but even of the universe."

-Archimedes, *The Sand Reckoner*

The Earth and the universe in which we live are riddled with mysteries. With this being the case, the list of questions continues to grow. How did the universe come into existence? From where in the universe are we looking up at the sky? Where is the Earth situated in the universe? Before answering these questions, let's begin by defining the universe.

The universe is a vast system that encompasses all celestial bodies, including Earth, as well as all forms of energy and matter. The universe includes all the planets, comets, galaxies, nebulae, and clouds of gas and dust that surround us.

According to the most widely accepted theory, this astoundingly beautiful and complex universe came into existence after the Big Bang. 13.8 billion years ago, our entire universe was small enough to fit inside one-billionth of a nuclear particle. And it grew quickly from a few centimeters to several hundred light-years. The universe expanded and then cooled, resulting in the formation of hydrogen and helium, which are among the lightest elements. Hydrogen and helium clouds condensed to create the first generation of stars, which, in turn, created the first small galaxies. When the first generation of stars died, the heavier elements that spread out in space enabled the formation of planets and, eventually, life.

The universe beyond the horizon is much larger than man can comprehend. Let's try to understand this massiveness through measurements from our lives: Fighter jets travel at approximately one kilometer per second. Even if you traveled at the speed of a fighter jet, it would take approximately a million years to reach Proxima Centauri, the closest star to the Sun. Considering the distances between galaxies and the Earth, this seemingly incomprehensible period is actually rather short. Besides, this is where the phrase "astronomical number" that we use in daily life comes from.

The unit we use in order to measure the universe is the light-year. One light-year is the distance that light travels in one year. A light-year equals 9,467,280,000.000 kilometers, which means light can travel at a speed of 299,792,458 kilometers per second. The speed of light is the speed limit of the universe. To put this magnitude into perspective, consider this: light will have completed seven laps around the Earth in the time it takes you to snap your fingers.

The observable universe, or the area that can be seen through telescopes, has a diameter of 13.8 billion light-years. I can sense that your head is spinning already. Once we add to that the universe, which we have yet to see or determine whether it is infinite or not, the matter truly transcends human comprehension. This, however, is true: the observable universe is only a tiny speck of the universe that we don't know!

The universe contains billions of galaxies. Galaxies, the universe's building blocks, are massive systems containing stars, star clusters[17], and nebulae[18]. The galaxy where our solar system is located is called "The Milky Way," but our Sun is just one of the approximately 300 billion stars in it! The billions of stars comprising the galaxy revolve around the galaxy center because of gravitational pull.

17 The stars are not evenly distributed throughout space. Stars born near each other comprise the "star clusters." In some cases, these stars remain together for billions of years. Clusters containing hundreds of stars are called "open clusters," whereas those containing thousands of stars are called "globular clusters."

18 The formation spots of baby stars or the remnants left by dead stars are, in the most general sense, called "nebulae."

When you look up at the sky, you may see some nebulous zones besides the Milky Way. This is Andromeda[19], which is yet another galaxy. Andromeda is an enormous galaxy that is 2.9 million light-years away from our Earth, but when taking into consideration the scale of the universe, it is right beside us. It is the largest object you can see with the naked eye; it spans 100,000 light-years from one end to the other.

Returning to the Milky Way, it is a barred spiral galaxy and is distinct from the other galaxies. While some galaxies are nearly spherical, the Milky Way is a flat disk with a bulge in the center. It appears as a bright belt on clear nights. Although the stars are sparsely scattered across the Milky Way, to us, it appears as if there are many stars in the same place. If we could see our galaxy, which has spiral arms, from above or below, we would see that it resembles a gigantic pinwheel.

Now, let's express the size of the Milky Way in light-years. It takes light 100,000 years to travel from one end of the Milky Way to the other. This means that the diameter of the Milky Way is 100,000 light-years. The Sun is 25,000 light-years away from the center of the Milky Way. At the center of the galaxy, there is a bulge made up of stars with a thickness of about 15,000 light-years. The Milky Way completes one full revolution around the center in 225 million years, which is one cosmic year.

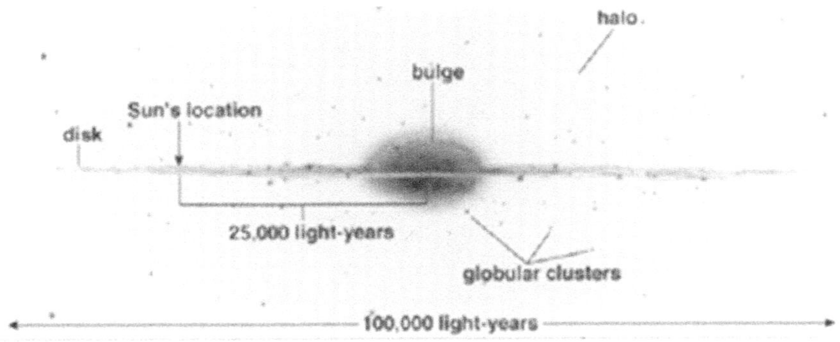

Image 10. The Milky Way

19 The area covered in space by Andromeda is about six times that of the full moon. However, it is so far away that, despite the 1000 billion stars it contains, only the central bulge of the galaxy can be seen with the naked eye.

As is seen, nothing in the universe is fixed. The Sun rotates around its own axis and revolves around the center of the solar system, while the Earth rotates around its own axis and revolves around the Sun. When we consider that our galaxy revolves around the galactic center and moves at certain speeds throughout the universe, I believe this is yet another reason for your head to spin!

The Milky Way is a member of the "local group," which contains more than 30 galaxies. Our local group is situated inside the Virgo Supercluster. A supercluster contains about 100 local groups. If someone asked you for your address in the universe, this is how you would respond: "I live on a planet in the solar system, which is situated in a galaxy called the Milky Way, which is situated in the local group inside the Virgo supercluster[20]." This address demonstrates that we occupy a very small space in the universe; there are millions of superclusters, countless local groups within each cluster, and countless galaxies within each local group. Now let's take a quick trip to see what's going on up there. Our first stop along the way is our solar system and its planets!

20 The largest galaxies in the Virgo supercluster are the Milky Way, Andromeda (M31), and the Pinwheel Galaxy.

III.
THE SOLAR SYSTEM

If we were to take attendance in the solar system, we would need to begin with the Sun, our star that comprises 99.9% of the system alone. Next in line would be the eight planets orbiting the Sun, as well as their satellites, asteroids, comets, meteorites, and cosmic dust. All of these components were created by the collapse of a primordial gas cloud and the subsequent collision of the particles created by the collapse. The pebble-sized pieces created the meteorites and larger celestial bodies. Finally, these celestial bodies gathered at the center to form planets and satellites.

First of all, let's address the eight planets that comprise the solar system. But first, let's have a look at the meaning of the word "planet." If we observe a planet for several nights in a row, we can see that its position changes with respect to the stars. This is why they are called "planets." In Greek, "planet" means "wandering." Five planets (Mercury, Venus, Mars, Jupiter, and Saturn) were known in the early eras, and Uranus, Neptune, and Pluto[21] were discovered through telescope observations after 1781.

All planets revolve around the Sun in their respective orbits. Only Venus, Uranus, and Pluto rotate clockwise around their own axis, while the other planets move counter-clockwise.

At first glance, the planets that make up the solar system can be divided into two groups. Mercury, Venus, Earth, and Mars, the first four planets closest to the Sun, are also known as "terrestrial planets." These planets are relatively small, and they are made up of rocky and metallic material. Jupiter, Saturn, Uranus, and Neptune belong to the second class of planets, which have lower densities and are farther away from the Sun. These planets are enormous gas giants with relatively small, solid cores.

---

21 Although Pluto was demoted from planet status in the world of astronomy, it is considered a planet in astrology.

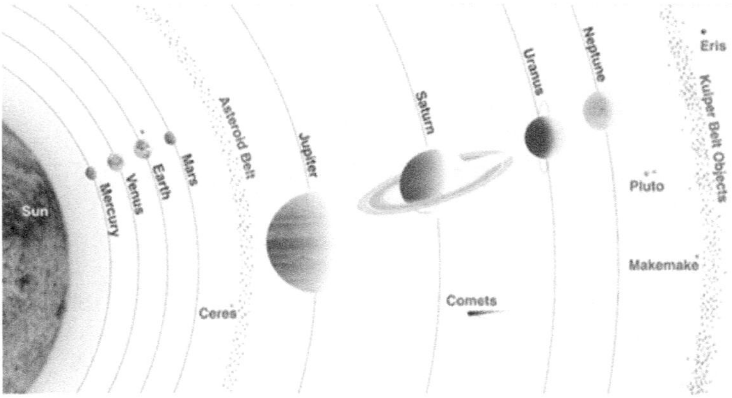

Image 11. The solar system

Pluto, which doesn't fit into either category, is the outermost planet in the solar system.

We can also split the planets into two categories based on their closeness to the Earth. Mercury and Venus, which are closer to the Sun than the Earth, are called inferior planets. Planets between Mars and Pluto, which are further from the Earth, are called superior planets. Inferior planets have phases, whereas exterior planets do not.

At this point, the planets' distances from the Sun are also worth noting. A planet's distance from another planet or from the Sun is measured in "Astronomic Units." The distance between the Earth and the Sun is one astronomical unit, or 1 AU. One astronomical unit equals 149.6 million kilometers. The following are the planets' distances from the Sun, from closest to farthest: Mercury 0.38 AU, Venus 0.72 AU, Mars 1.52 AU, Jupiter 5.2 AU, Saturn 9.53 AU, Uranus 19.19 AU, Neptune 30.06 AU, and Pluto 39.53 AU.

The planets revolve around the Sun in nearly the same plane. The orbit of Pluto makes a 17,15° incline to the plane of the ecliptic, which is the highest in comparison to other planets. (For instance, this incline is 7° for Mercury, 3.39° for Venus, and 0,77° for Uranus.) Mars, Saturn, and Neptune, all of which have axial tilts close to that of the Earth[22], have

22 The angle between a planet's rotational axis and its orbit aro-und the Sun is called the planet's "axial tilt." Planets have varying axial tilts with respect to their orbital planes. (See Image 13.)

seasons like the Earth. Mercury and Jupiter, on the other hand, do not have seasons.

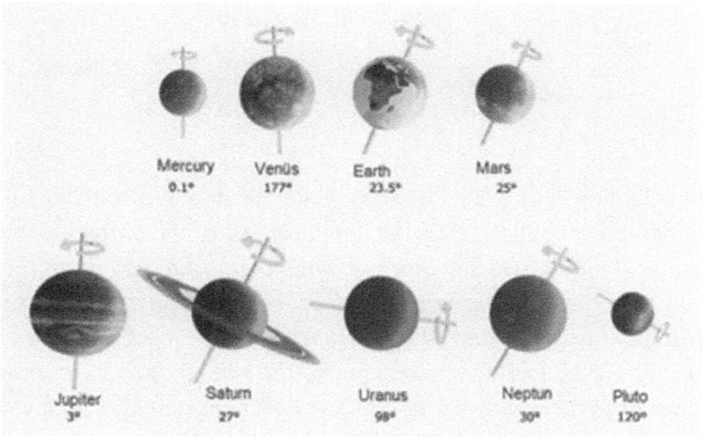

Image 12. The axial tilts of the planets

The first zone between Mars and Jupiter is called the "Main Asteroid Belt." This belt contains countless rocks, metallic pieces, and mostly potato-shaped asteroids. This is the zone in science fiction movies through which spacecraft pass, where gigantic celestial bodies reside, and which separates the planets from one another.

In 1992, a reservoir full of icy particles was discovered beyond Neptune, accompanying Pluto; an extremely cold belt assumed to have come into existence in the early phases of the solar system. Consisting of dirty snowballs of various sorts and sizes, this belt is called "the Kuiper Belt." Substances made up of the conglomerated dust of exploded stars are located here. Comets come from beyond the Kuiper belt.

THE SUN

"There is a sun, a light that for want of another word I can only call yellow, pale sulphur yellow, pale golden citron. How lovely yellow is!"

-Vincent Van Gogh

When Copernicus was just a student, he had written the following words about the Sun: *"...of all the bodies in the universe the most excellent is the Sun, whose whole essence is nothing else than the purest light (...) which singly and alone is the producer, conserver, and warmer of all things; it is a fountain of light, rich in fruitful heat, most fair, limpid, and pure to the sight (...) called king of the planets for its motion, heart of the world for his power, its eye for his beauty, and which alone we should judge worthy of the Most High God, should he be pleased with a material domicile and choose a place in which to dwell with the blessed angels."* Considering these words with which Copernicus has expressed his admiration for the Sun, it is no wonder why people of the ancient era have attributed chief god status to it!

The Sun, the most prominent star in the sky, is only one of billions of stars in the Milky Way. Compared to giant stars like Betelgeuse and Antares, the Sun, our source of life, is classified as a dwarf star. Although it may appear small and insignificant in comparison to the other stars, it is critical to our survival. Regardless of the distance[23] between us and the Sun, it is our source of heat, light, and life.

The layers comprising the Sun can be analyzed in four parts. These are the core, the photosphere (the light sphere), chromosphere (the color sphere), and corona (the crown sphere). The photosphere, chromosphere[24], and corona may be referred to as "the atmosphere of the Sun."

Our Sun is about 4.5 billion years old. Despite its ostensibly long history, it is not a first-generation star. This blazing sphere, which originated from

23 The distance between the Sun and the Earth is around 150,000,000 kilometers, which is roughly equal to 3700 laps around the Earth. Planets revolve around the Sun in an elliptical orbit, which means their distances from the Sun fluctuate. The average value is calculated from the aphelion and perihelion points.

24 We can only see the chromosphere during a solar eclipse.

the substances emitted by former stars during supernova[25] explosions, draws its energy[26] from nuclear reactions that convert the hydrogen in its core to helium. Its mean surface temperature is 5500°C, and its mean core temperature is 16,000,000°C. The extremely high temperature of its core causes hydrogen to be converted into helium and photons to be emitted. Photons or beams created at the core of the Sun take 10 million years to reach the surface, whereas the photons on the surface of the Sun take 8 minutes and 20 seconds to reach our Earth. As a matter of fact, the light we see when we look at the Sun left the Sun's surface 8 minutes and 20 seconds ago, but was created 10 million years ago![27]

So, how big is the Sun? Let's try to explain this in terms we use in daily life: let's say that the Earth is the size of a single lentil. If we line up 109 of those lentils side by side, their total length equals the diameter[28] of the Sun, which is nearly the size of a basketball. Or, if we compare the volume of the Sun to that of the Earth, we can say that a million Earths could fit inside the Sun.

We assume that the Sun is immobile, but like all other members of the solar system, the Sun revolves around the center of mass of the solar system. It also revolves around the center of our galaxy, completing one full revolution in about 250 million years. Another motion of the Sun is its rotation around its own axis, as evidenced by sunspots. Galileo noticed these spots while observing the Sun.[29] He assumed they were dark because they were holes in the Sun's surface. We now know that these spots are caused by magnetic activity that inhibits energy from coming

---

25 Supernovas are the violent explosions of stars that have depleted their energy.

26 In just one second, the Sun generates enough energy to meet the energy requirements of the entire Earth for half a billion years.

27 It takes time for light to reach us. The light from the nearest star takes years, whereas the light from other stars may take centuries. Some stars are so far from the Earth that it takes centuries for their light to reach us. In fact, by the time the light from some stars reaches us, the stars themselves are long dead. Therefore, the light coming from a star that is billions of light-years away is, in fact, that star's appearance from billions of years ago, or in other words, its past.

28 The diameter of the Sun is 1,392,000 kilometers.

29 Galileo wasn't the first to notice sunspots. Chinese astronomers observed sunspots much earlier in history. The earliest known records of sunspots are those made by a Chinese astronomer named Gan De around the 360s BC. We know that around the 300s BC, the Greek scholar Theophrastus kept records of sunspots. These observations continued through-out Europe and the Orient in the common era as well. So, we can consider Galileo to be the first person to make sense of sunspots.

out. The number of sunspots varies with changes in the Sun's magnetic field. This process is repeated every eleven years. Small sunspots can survive for a week, while larger ones can survive for anywhere from a month to a month and a half. Sunspots cover one to two millionths of the Sun's surface.

Galileo discovered that the Sun rotated around its own axis and calculated the rotational period of the Sun by tracking sunspots. Because the Sun is a gas mass, not all of the planet rotates at the same rate. According to an observer on Earth, it completes one full revolution in about 24 days around the equator and about 30 days in the zones near the poles. It revolves around itself at a speed of 220 kilometers per hour.

Finally, let’s talk about solar winds. You must have heard of the auroras, particularly the Aurora Borealis, also known as the polar lights. Auroras, which can be seen from Earth and put on a colorful show, are linked to changes in the number and surface area of sunspots. This fluctuation causes solar activity. In turn, these activities cause solar winds to spread out into the interplanetary environment. When these winds, which may reach speeds of up to 400 kilometers per second, enter the Earth's magnetic field, they cause light to be emitted. Auroras are the products of such magnetic storms.

As long as the Sun emits light, life on Earth will continue. However, 4.5 billion years later, the Sun will have depleted all of its hydrogen, and as helium atoms coalesce and begin to turn into carbon, its heat will increase even more. It will expand to cool itself before contracting once more. It will then explode and become a white dwarf[30], shining more dimly.

30 Small stars with a surface temperature of approximately 100,000° and a diameter between 1000 and 2000 kilometers are called "white dwarfs." It is the last phase in the evolution of Sun-like stars.

THE MOON

"GALILEO: How do you explain those luminous spots?
SAGREDO: It cannot be.
GALILEO: But it is. They are mountains.
SAGREDO: On a star?
GALILEO: Giant mountains. Whose summits are gilded by the rising sun, whilst all around night still covers their slopes. You see the light descending from the topmost peaks into the valleys.
SAGREDO: But that contradicts all astronomy for the last two thousand years."

-Bertolt Brecht, the definition of the Moon in his play titled *Life Of Galileo*

About 4 billion years ago, our planet collided with another planet the size of Mars. After this collision, the pieces that broke off of the Earth scattered across space. All of these leftover pieces eventually came together to form the Moon, our silvery satellite.

The Moon is the second brightest object in the sky after the Sun. It is not, however, a source of light in and of itself; rather, it reflects the light beams it receives from the Sun. This is due to its close proximity to the Earth. The Moon is the Earth's only natural satellite.

The distance between our Earth and the Moon is 384,500 kilometers. This might not be considered an astronomical measurement, as it is a distance that a human can travel in a lifetime. Consider this: the distance between Izmir and Istanbul is approximately 550 kilometers. So, a round trip between the two cities would be 1100 kilometers in total. If you travel from Izmir to Istanbul and then back to Izmir 350 times, you will have traveled the distance between the Earth and the Moon. At 3476 kilometers, the Moon's diameter is also of a conceivable scale. This is about one-fourth the diameter of the Earth.

The Moon has little to no atmosphere. Therefore, the mean temperature difference between day and night can reach 100°C. According to the most recent data, the Moon's maximum temperature during the day was detected to be +117°C, and its minimum temperature during the night, −117°C.

The Moon is the only terrestrial body that has been visited in the history of humanity. In 1952, the Luna-2 satel-lite landed on the surface of the Moon, revealing two major surface features: the highlands, which are craterized ridges, and the maria, which are relatively flatter formations. You can observe the lunar maria with the naked eye. Lunar maria are the parts of the Moon that appear dark. People used to think these were seas, which was why they called them "maria."[31] Lunar maria are mostly found on the visible side of the Moon, and they comprise about 31% of the Moon's surface. Lunar maria are known by a variety of names, including Mare Tranquillitatis (Sea of Tranquility), Mare Nubium (Sea of Clouds), Mare Crisium (Sea of Crises), and Lacus Somniorum (Lake of Dreams). These names are quite in line with the astrological symbolism of the Moon.

Since our silvery satellite doesn't have a thick, protective atmosphere like the Earth, its surface has been subjected to many meteorite impacts. This is why craters, which are among the most prominent features on the Moon, can be found almost all over the Moon. At least 300,000 of the craters are larger than one kilometer in diameter. The larger craters' diameters can exceed 100 kilometers. There are mountains on the Moon's surface as well, though they are not as prominent. Lunar mountains are the highlands of the Moon, and they appear brighter because of their material composition.

The Moon moves around more compared to the stars and planets. It has three motions: rotating around itself, revolving around the Earth, and revolving around the Sun along with the Earth. The Earth revolves around the Sun in the same direction as the Moon. It takes 29 days, 12 hours, 44 minutes, and 2.8 seconds for the Moon to return to its same appearance. This equals the time interval between two full moons. During its diurnal motion, the Moon moves by 13° 11′ in relation to the stars.

To an observer on Earth, the Moon's rotation period around its own axis is the same as its revolution period around the Earth—27 days, 7 hours, 43 minutes, and 11.6 seconds. This is why the Moon always presents us the same side.

---

31 In Latin, *maria* means "seas" (plural), whereas *mare* means "sea" (singular). The maria take up approximately 16% of the Moon's surface. These maria, which were created as a result of collisions, are more abundant on the side of the Moon that faces the Earth.

The colloquial phrase "the dark side of the Moon" is misleading since it implies that the "dark side of the Moon" receives no light at all. However, the back of the Moon receives light from the Sun as well.[32] Satellite images show that the back of the Moon has more craters. This is because the back of the Moon is not under the protection of the Earth and is hence more vulnerable to meteorite collisions.

The Moon and Earth's gravitational influence on one another cause tidal events. As the Earth rotates around its own axis faster than the Moon moves in its orbit, these bulges occur twice a day. Tidal events lead to a decrease in the Earth's rotational speed; this slows down the rotation of the Earth by about 2 milliseconds. This decrease in the Earth's speed necessitates the Moon to move a little further away from us; the Moon drifts away from the Earth by about 3.8 centimeters per year.

MERCURY

Mercury is a planet whose presence has been known since the Sumerians' time. There are observational records of Mercury dating back to 264 BC. The Greeks named Mercury differently depending on whether it was east or west of the Sun; they called it Apollo when it was a morning star and Mercury when it was an evening star.

Mercury is the closest planet to the Sun. Thus, it is even brighter than Sirius, which is among the biggest stars in the sky. It is, however, so close to the Sun[33] that it is rarely visible with the naked eye. To see it, one must look at the right place at the right time. It can only be seen in the west just after sunset and in the east, at a point very close to the horizon just before sunset.

32 We can see 41% of the Moon's surface at all times and 18% of it on occasion. We never see the remaining 41%. This is attributable to the Moon's orbit being 5% inclined to that of the Earth, as well as irregularities in the Moon's orbital motion.

33 Mercury's distance from the Sun is 46 million kilometers at perihelion and 70 million kilometers at aphelion. Its mean distance from the Sun is about 58 million kilometers.

Mercury's quick motion around the Sun has caused it to lose its perfectly round shape. It is the planet with the highest eccentricity. Its mass is 0.38 times that of the Earth. So, if you weigh 60 kilograms (132 pounds) on Earth, you would weigh 15 kilograms (33 pounds) on Mercury. Mercury has a diameter of 4870 kilometers, which is comparable to the diameter of our Moon (3472 kilometers).

Mercury's orbit passes between the Sun and the Earth. Therefore, the planet has phases, like those of the Moon. Mercury has two distinct sides: the side that faces the Sun gets quite hot, reaching +440°C. The other side, on the other hand, is quite cold; temperatures on this side can drop as low as −280°C. We always see one side of Mercury, similar to how we always see one side of the Moon.

If we were to set foot on Mercury, we might have thought that we had set foot on the Moon, as Mercury's surface is also studded with craters, like that of the Moon. In terms of physical structure, it is thought to be similar to the Moon. The information we have on Mercury is based on observations made by Mariner 10, a spacecraft that flew by the planet between 1974 and 1975. Mercury differs from the Moon in that it lacks the vast plains known as lunar maria on its surface; there are only intercrater plains. It is slightly larger than the Moon. Another similarity between the Moon and Mercury is their water vapor content. In 2009, Mercury's north pole craters, which remained in the dark, were probed, and water ice was discovered. We know that the Moon also has water vapor at its poles. This information can be interpreted as an astronomical explanation for the Moon-Mercury connection emphasized in classical astrology.

The orbit of Mercury is more ecliptic than circular. Therefore, the greatest elongations for Mercury range between 18° and 28°. Its orbit is inclined by about 7° with respect to the ecliptic. Its orbital speed is 58 kilometers per second at perihelion and 38 kilometers per second at aphelion.

Mercury's axial inclination of 0.1° is negligible in comparison to the Earth's axial inclination of 23.5°; thus, Mercury spins almost upright in relation to its orbital plane. The planet completes one full revolution around the Sun in 88 days, which means that one Mercury year equals 88 Earth days. Its rotation around itself, on the other hand, is slow; one Mercury day equals 58 Earth days.

VENUS

Venus, named after the Roman goddess of love, is the third brightest visible object in the sky after the Sun and the Moon.[34] The Greeks considered Venus to be two separate celestial bodies (as they did Mercury) and referred to the planet's appearance in the morning as the "morning star" (Phosphorus) and its appearance in the evening as the "evening star" (Hesperus).[35] It is seen at two different times a day because its orbit is closer to the Sun compared to that of the Earth. Venus is the second closest planet to the Sun; its distance from the Sun is twice that of Mercury.[36]

In the solar system, Venus has an almost perfectly circular orbit. It revolves in the opposite direction than the Earth or Mars—it revolves from east to west.[37] Consequently, if we could observe from Venus, the Sun would be rising from the west.

Venus, which is roughly the same size as our Earth, is also the planet that is the most similar to it. In fact, up until 50 years ago, Venus was believed to be the Earth's tropical sister planet. Venera 9, which sent the first images from Venus's surface to the Earth in 1975, proved that Venus was entirely different from what was previously assumed. Although it was similar to Earth in terms of mass, size, and even chemical composition and density, the two planets had fundamental differences. Venus was a scorching hot planet. It had a surface temperature of over 500°C. This discovery debunked the theory that Venus was covered in vast seas. Furthermore, the planet's high atmospheric pressure—90 atm, to be exact—made it certain that no liquid water would be discovered on Venus.

---

34 Venus is 15 times brighter than Sirius, which is the brightest star in the sky, and 6 times brighter than Jupiter.

35 In Turkish mythology, Venus also takes the name "Zühre," the protector of shepherds. Considering that it shines as a guiding light, it can accompany shepherds both in the morning and in the evening. Venus guides people in orienting themselves; if it appears in the morning, it points westward, and if it appears in the evening, it points eastward.

36 Mercury's distance from the Sun is 108.2 million kilometers (0.72 AU).

37 Venus rotates backwards because of the Sun's powerful gravitational pull, which, in turn, causes strong atmospheric tides.

If we were on Venus, we wouldn't be able to see the Sun or the Earth under any condition, because the planet's atmosphere, which is thick and high in carbon dioxide (about 95%), would not allow it. This is also why Venus appears so bright from the Earth: the kilometers-thick acid clouds in its atmosphere. These acid clouds create a green-house effect. This is what gives Venus its reddish appearance.

The surface of the reddish-orange planet Venus has volcanic structures of various sizes. With over 1600 volcanoes, the planet has a hellish appearance. Over 80% of its surface consists of volcanic plains formed by lava outflows that have cooled down and become solid. Although it isn't clear if these volcanic activities are still ongoing, the planet has all the conditions it needs to be volcanically active.

Since Venus is an inferior planet, it has phases such as "full" or "crescent."[38] In 1610, Italian astronomer Galileo noticed the phases of Venus when observing the planet through a telescope. He realized that these were caused by Venus's orbital motion around the Sun as well as the fact that Venus reflects sunlight.

Venus's revolution period around the Sun is 225 days. Although its rotation period around itself hasn't been determined precisely, it was found to be slightly longer than 243 days. This means that one Venus day is longer than one Venus year (the time it takes for Venus to complete one lap around the Sun).[39] Venus's orbit is more ecliptic than circular, meaning it is oblate. This is why the greatest elongations for Venus range between 47° and 48°.

---

38 When Venus is at its closest to the Earth, the side of it facing the Earth goes completely dark. When it is at its "full" phase, which is when its illuminated side is entirely visible, it is at its farthest position and its smallest apparent size. It is not bright enough. The planet sends its observers the most light when the highest amount of its surface is visible, at 30% brightness (between the crescent and half phases).

39 It is believed that Venus doesn't have a magnetosphere since its rotation around itself is quite slow.

MARS

Although Mars is similar to our Earth in many ways, the two planets differ significantly in others. Let's start with the similarities: Mars has an axial tilt of about 25°, which is very close to that of the Earth. Therefore, it is summer in the southern hemisphere of Mars when the planet is at perihelion, just as it is on Earth. Another similarity between Mars and our planet is their rotation periods: at 24 hours, 37 minutes, and 23 seconds, the rotation period of Mars is nearly the same as that of the Earth. Mars takes 687 days to complete one full revolution around the Sun, so one Mars year equals 687 Earth days.

Physically, however, they are worlds apart, so to speak: Mars is a planet on which volcanoes, canyons, cliffs, plains, and caves prevail. It also has the tallest mountain in the solar system, called Olympus Mons (Latin for Mount Olympus), which stands approximately 28 kilometers tall.[40] It is similar to the Ihlara Valley in Turkey, but it has much larger canyons (4000 kilometers in length and 2 to 7 kilometers in depth), which gives its surface an intriguing, enormous appearance. Contrary to popular belief, Mars, which has inspired many science fiction stories, is a volcanic wasteland! It is a much colder planet than our Earth because of its relative distance from the Sun. Its mean surface temperature is −55°C, although it drops as low as −123°C during winter.

Mars's distance from the Sun is 227.94 million kilometers. This is a little over half the distance between the Earth and the Sun. Mars's diameter is 6794 kilometers, which falls somewhere between the diameters of the Earth and the Moon; it is almost half the diameter of the Earth and twice the diameter of the Moon. Its mass is about 11% of that of the Earth.

Mars, also known as the "red planet," is characterized by a reddish rust color that covers more than half of its surface. Mars' red appearance is caused by the oxidation of iron in the planet's rocks, soil, and dust in its atmosphere. In other words, the planet's red appearance is caused by rust formation. The other parts are covered in black and blue spots. These spots are indicators that the planet may have an atmosphere. Research also shows that Mars has a very thin atmosphere. The thinness of its atmosphere allows the planet to be seen from Earth. Carbon dioxide accounts for a sizable portion of the atmosphere—95%, to be exact.

---

40 Olympus Mons is three times the height of Mount Everest.

Mars has two natural satellites: Phobos[41] and Deimos.[42] Both satellites' surfaces are covered in craters. These amorphous satellites, kilometers wide in diameter, are believed to be former asteroids captured by Mars. If we could observe Phobos and Deimos from Mars, Phobos would appear to us to be one-third the size of the Moon as seen from Earth, and Deimos, one-ninth the size. Because of tidal effects, both satellites only turn one side toward Mars, similar to how the Moon always has one side turned toward us.

Phobos' orbit gradually shrinks as it revolves around Mars faster than Mars rotates around itself. Phobos is expected to crash into Mars in a hundred million years. Deimos' orbit gradually expands as it is quite far from Mars.

JUPITER

Moving beyond the main asteroid belt, we see Jupiter, the giant of the solar system. Jupiter is the largest planet in the solar system. Its colossal size is also reflected in mythology; the planet is named after the god of gods of Roman mythology. Jupiter can hold thousands of Earths. Its mass is 318 times that of the Earth. To put this into perspective, I think it would suffice to say that Jupiter's mass is more than 2.5 times that of all the other planets in the solar system combined. It is by far the heaviest planet in the solar system. Had its mass been 80 times greater, it could have been a star. Its other phenomenon of greatness is apparent in its magnetic field. It has the largest magnetosphere of all the planets in the solar system. Its magnetic field is 19,000 times stronger than that of the Earth.

Jupiter's volume is greater than the combined volumes of all the other planets. It has the largest diameter that a gas planet can have: 143,000 kilometers. This is why you may hear some people say, "The solar system consists of Jupiter and assorted debris." This isn't inaccurate at all. The fact that it is so large makes it the brightest object in the sky after Venus. Its largeness doesn't slow down its rotation—on the contrary, it causes

41 Since Phobos orbits Mars below the synchronous orbit radius, to an observer on Mars, it rises and sets about twice a day.

42 The names of these satellites mean "fear" and "panic," respectively.

it to rotate faster. Its size is also the reason for its ellipse shape. At 9.92 hours, one Jupiter day is shorter than that of any other planet in the solar system. It completes one full revolution around the Sun in 11.86 years.

Jupiter has no solid surface because it is made of gas; 90% of it is hydrogen and 10% is helium. Deeper inside the planet, compressed liquid versions of these elements are found. The most striking feature in the upper layer of its atmosphere is its clouds. Jupiter is a colorful place, with distinctive bands and storms swirling around in bright colors. The lightning strikes there are hundreds of times more violent than those on Earth. The hailstones are as large as footballs. Strong winds that can reach a speed of 500 kilometers per hour are observed in its atmosphere.

The most striking atmospheric feature of Jupiter is the "Great Red Spot." The likes of this feature, which is large enough to contain two Earths, are also found on Saturn and Neptune. In the upper levels of the atmosphere, there are fine, faint, and dark rings that cannot even be observed through the most powerful telescope. However, these rings aren't clearly visible like the rings of Saturn. This is because of the rings' low mass and their inability to reflect sunlight well.

With its 63 known satellites, Jupiter has the most crowded satellite family in the sky. Its four major satellites, discovered by Galileo in 1610, are Io, Europa, Ganymede, and Callisto, and they are collectively known as the "Galilean moons." These satellites get their names from four of Zeus's lovers. Since the other satellites of Jupiter have diameters of less than 320 kilometers, they cannot be observed through a telescope.

The nature of Jupiter's satellites is just as colorful as the planet's appearance. Ganymede is covered in an icy shell, whereas Ganymede has lava flows. Ganymede is larger than Mercury; it is the largest satellite of Jupiter as well as of the solar system. Io is the satellite that is closest to Jupiter. For the last 400 years, we had been assuming that Io, like our Moon, was a dead place. However, when the first spacecraft flew by Jupiter in the late 1970s, we realized that our assumption was far from the truth. Io is the most volcanically active body in the solar system.[43] Just one of Io's lava flows alone generates more heat than all the volcanos on Earth combined.

43 Io's gravitational pull and Jupiter's extraordinarily powerful gravitational pull lead to friction inside the satellite, which, in turn, causes the volcanic activities on Io.

In astrology, Jupiter is known as the "greater benefic." By protecting the Earth from many asteroids and comets, it affirms its position as the Earth's insurance in astronomy as well. Among all the planets, Jupiter has the strongest magnetic field. Its gravitational field has a significant impact on interplanetary wandering material. Since it has such a large mass, many asteroids and comets face the risk of colliding into Jupiter first. Looking at the situation from a different angle, Jupiter's massiveness may also alter the trajectory of an asteroid that would probably never collide with the Earth otherwise, thus causing it to pose a danger to us. But in any case, Jupiter is, overall, a true protector of our Earth.

SATURN

Saturn is the second largest planet in the solar system. Its distance from the Sun is approximately 1,429,400,000 kilometers (9.45 AU). This is more than twice the distance from the Sun to Jupiter.

Although Saturn could swallow 765 Earths in terms of volume, its density is less than that of water. So, if there was a suitable ocean in which we could drop Saturn, it would float. Saturn's mass is 94 times that of the Earth.

Saturn can be observed with the naked eye at night, and it is almost as bright as Jupiter. The atmosphere of the large gas planet is 75% hydrogen and 25% helium, similar to the atmosphere of Jupiter. The yellowish color of the planet almost makes it seem as if it were made of lead. When observed through a telescope, Saturn's most prominent feature is its rings. As a matter of fact, the other four giant planets have rings as well. Saturn's rings stand out because of their very bright and large appearances.

Here is how Galileo described his observations of the rings of Saturn, which are known as the marvels of the solar system:

"(...) I have found out with the most intense surprise that the planet Saturn is not merely one single star, but three stars very close together, so much so that they are all but in contact with one another. (...) The middle star of the three is by far greater than the two on either side.

(...) if you should use a glass which multiplies a surface more than a thousand times, there will appear very distinctly three orbs, almost touching one another; and they will be thought to be not further apart than the breadth of a very fine and scarcely visible thread."

Observations showed that Saturn was surrounded by 3 separate rings, as Galileo had also stated. However, in the 1970s, the number of rings went up to 6. To an observer on Earth, these rings appear continuous, with no gaps or breaks. The data taken from the Voyager probes established that each one of these rings consisted of small, concentrical ring-lets. Thus, the number of rings exceeded a thousand.

As a matter of fact, these rings, which are parallel to the planet's equator, move in different orbits. Each ring is made up of small pieces ranging in size from a few centimeters to a few meters. Saturn's rings are so visible because they are made of water ice, which is reflective. The particles that make up the rings are thought to have formed as a result of a larger celestial body colliding with Saturn. The origin of these particles is estimated to be about 500 million years old. This tells us that these rings do not have the same origin as our 4.5-billion-years-old solar system!

The rings spread as far as 480.000 kilometers from the center of the planet. So, Saturn's rings have diameters five times that of the planet itself. For the most part, the thickness of its rings doesn't exceed several hundred meters. Because of the planet's prominent axial tilt with respect to its orbital plane, both sides of its rings are visible.

Saturn completes one full revolution around the Sun in 29.46 years and one full rotation around itself in 10.25 hours. Saturn has the most crowded family of satellites after Jupiter. It has at least 48 known satellites. The largest among these—and the only one that is known to have methane in its atmosphere—is Titan. Its diameter is greater than that of the Moon. Some of Saturn's satellites are larger than Mercury, and some are smaller. Mimas, Enceladus, Tethys, Dione, Rhea, Titan, Hyperion, Iapetus, and Phoebe are its largest satellites, in order of proximity to the planet.

URANUS

Until the latter half of the eighteenth century, the only planets whose existence was known were the five planets that were visible to the naked eye. On March 13, 1781, an amateur observer called William Herschel was observing the constellation Gemini with a telescope that he had made himself when he came across a different celestial body. At first, he assumed that this object was a comet. Upon the observations and calculations that he conducted later, he realized that he had, in fact, discovered the giant gas planet Uranus, which would be a milestone in the history of astronomy.

The process of naming Uranus was just as interesting as its abrupt discovery. In truth, Uranus had been observed for a long time, but it was disregarded because it was assumed to be a star. The first Astronomer Loyal, John Flam-steed, had seen Uranus six times in total between the years of 1690 and 1715. Assuming it was just a regular star, he had given it a stellar number (34 Tauri) and hadn't made much else of it. When Herschel discovered Uranus, he named the planet "Georgium Sidus" after King George II of England. Later on, however, mythology had the last word again: the planet was named after Uranus (Ouranos), the god of the skies.

Uranus, covered in blue haze, consists of immense gases—hydrogen, helium, and methane—which give the planet its color. The planet's blue appearance is caused by the methane in the upper layers of its atmosphere that absorbs red light. It has a frozen and hazy atmosphere with a temperature that drops as low as −215°C. Uranus has a mass about 15 times that of Earth and a volume 100 times that of Earth.

If we were to take a closer look at Uranus, we could say that it was a motionless planet, as calm and serene as a pool. But looks can be deceiving. Despite its appearance, Uranus is a planet with many interesting qualities. Its atmosphere changes depending on the seasons, its magnetic field is lopsided, its rings are unlike those of any other planet, and it rotates on its side.

Uranus consists of ten dark and clear-cut rings as sharp as blades. These rings are eccentric, and they are situated quite far apart from each other. Its magnetic field, on the other hand, is truly odd and extremely volatile. Its magnetic axis is tilted at 60° from its axis of rotation, and it does not pass through the center of the sphere. Thus, the pole that we call the north pole of rotation is the magnetic south pole, and polar lights are mostly seen around its equator rather than its poles.

Another intriguing quality of Uranus is that it is tilted. This is believed to be caused by the gravity of Uranus's enormous neighboring planets or a collision with another planet. Imagine the chaos that would ensue if we tilted the Earth on its side! Uranus has an axial tilt of 97.9°, which is over 90°. While most planets in the solar system rotate almost perpendicular to the ecliptic on their own axes, the rotational axis of Uranus is nearly parallel to the ecliptic, like a gigantic target board. This indicates that Uranus has retrograde rotation. Many questions about this motion are yet to be answered by scientists.

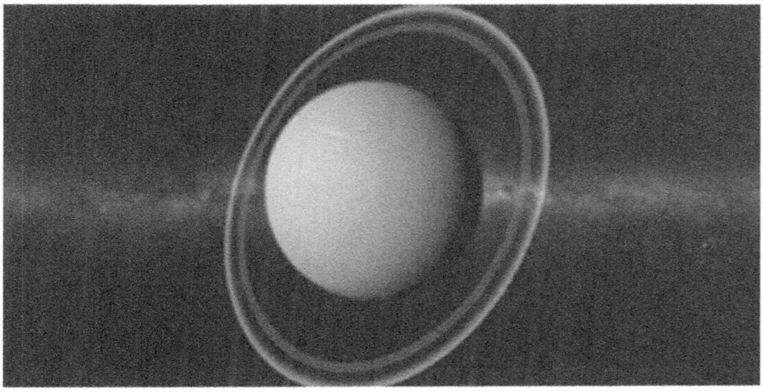

Image 13. Uranus

Uranus, which can serve as a cosmical source of many jokes with its peculiar qualities, is larger than Neptune in diameter but smaller than it in mass. Its distance from the Sun is over twice that of Saturn (19 AU=2,870,990,000 kilometers). It completes one full revolution around the Sun in 84 years and one full rotation around itself in 17.24 hours.

With its 27 known satellites, Uranus is the third planet with the most satellites, the first two being Jupiter and Saturn. Five of its biggest satellites are Titania, Oberon, Umbriel, Ariel, and Miranda. Their diameters range from 500 to 1600 kilometers. Five of its satellites move steadily in their respective orbits, whereas the others do not; if they get too close to Uranus, they may disappear.

NEPTUNE

Neptune, the last of the giant planets, is found in the depths of the solar system, 21 AU[44] (1495978286.77 kilometers) ahead of Uranus.

The first person to discover Neptune was Galileo. The neighboring star that Galileo had seen while observing the four satellites of Jupiter was, in fact, Neptune. Galileo hadn't realized that this new celestial body was Neptune.

Neptune's existence was known even when it couldn't be observed. English astronomer John Couch Adams and French astronomer Urbain Jean Joseph Leverrier conducted researches in order to explain the irregularities in the orbit of Uranus, in 1845 and 1846, respectively. After 40 years of location observations, it was understood that these irregularities were caused by the gravitational pull of an unknown planet in the solar system. Neptune was discovered on paper because it was impossible to observe it with the telescopes that were available at the time. In the year 1846, when there were no mechanical or electronic calculations, it was understood that this celestial body was a planet and its coordinates were determined, all based on a small number of highly flawed observations. When it was finally observed, it became clear that the planet's location was about 1° off from what was previously assumed.

If we could move in the depths of space, we would notice the beauty of Neptune, Uranus's icy sister, right away. With its blue-green atmosphere and white clouds, Neptune's appearance is proof of its beauty. This color tells us that there is a large amount of hydrogen in Neptune's atmosphere. With a surface temperature of −210°C, it is an extremely cold planet.

One of Neptune's mysteries is its spots. It has five large spots on its surface, three of which are bright and the other two are dark. The bright spots look like ink dropped in swirling water. The largest of its spots is called the "Great Dark Spot." Because of the violent storms in the solar system, these spots appear as if they will fall apart at any moment. The locations and sizes of these spots change over the course of 5 years, albeit slightly. When the Hubble space telescope turned toward Neptune, one of its spots vanished. These spots suggest that everything is immobilized by frost and that there are no atmospheric flows.

---

44 Neptune's distance from the Sun is 30.06 AU.

Neptune's distance from the Sun is about 4500 million kilometers, which is over 30 times that of the Earth. Neptune is also about 4 times as large as our Earth. It has a diameter of 50,000 kilometers. It has 65 times the volume and 17 times the mass of the Earth.

It completes one full revolution around the Sun in 164.8 years and one full rotation around itself in 16.1 hours. It has two of the largest satellites in the solar system. These satellites are called Triton and Nereid. Triton is larger than the Moon. The orbits of these two satellites are in the opposite direction to Neptune's orbit.

Neptune is a border planet in the belt called "Kuiper's Belt." Pluto entered Neptune's orbit in 1979 and remained there until 1999, making Neptune the outermost planet in the solar system during this time.

PLUTO

The story of Pluto's discovery is similar to that of Neptune. Percival Lowell, who believed that there should be a planet beyond Neptune, spent a long time obsessively searching for Pluto in the skies. Despite his best efforts, he was unable to produce any results because Pluto was smaller and fainter than expected. This was why Lowell had missed Pluto. Upon the death of Lowell in 1916, the search for Pluto was cast aside until 1926, when an astronomer named Cyle W. Tombaugh took over the search mission for Pluto. For four years, he scoured half a million stars for the slightest movement. On February 18, 1930, on the birthday of Lowell, who dedicated years of his life to discovering Pluto, he discovered that the planet kept disappearing and reappearing; Pluto had moved. Pluto was thus discovered after a long and exhausting search.

The naming of the planet was as eventful as its discovery process. Eventually, the planet was named Pluto, as suggested by an 11-year-old girl. This was an apt decision, since the planet was as cold and dark as the Roman god after whom it was named.

At a meeting on August 24th, 2006, in Prague, Pluto was demoted from planet status and was re-classified as a dwarf planet[45] instead. Since it is so far away, it appears as a very faint dot of light, even when observed through large telescopes. Pluto has a thick but sparse atmosphere that essentially consists of nitrogen and a small amount of carbon monoxide. According to observations, its diameter is likely to be around 2390 kilometers. It is one-fourth the size of the Moon.

For a long time, Pluto's first known satellite was Charon. Charon is half the size of Pluto (1186 kilometers). Pluto and Charon revolve as if they are locked on each other, and Charon always keeps the same side facing toward Pluto, similar to how the Moon always keeps the same side facing the Earth. This satellite has a very thin atmosphere. Some astronomers claim that Pluto and Charon were formerly Neptune's satellites but were detached from it after a collision. In 2005, it was revealed that Pluto also had two small satellites. In 2006, these satellites were named Hydra and Nix.

Pluto's density is more than twice that of water. Its mass is approximately 0.6% of that of the Earth. This means that someone who weighs 70 kilos on Earth would weigh about 4.5 kilos on Pluto.

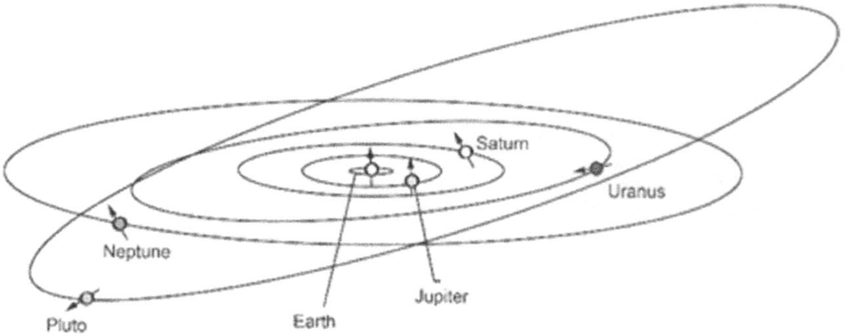

Figure 5. The orbit of Pluto (shown from Jupiter)

45 The International Astronomical Union declared in 2006 that a celestial body must meet the following criteria to be considered a planet: it must be in orbit around the Sun, it must have a spherical structure and a specific mass, and it must have enough gravity to pull in and expel the materials around it. The most widely known dwarf planets in the solar system are Ceres, Pluto, Eris, Haumea, and Makemake.

Pluto has the greatest orbital inclination of any planet in the solar system. At perihelion, it can get closer to the Sun than Neptune. For instance, between the years of 1979 and 1999, the distance between the Sun and Pluto was less than the distance between the Sun and Neptune. One might wonder if these two planets might collide one day. However, this is no cause for concern, as the angle between the planes of the two planets is over 17°. Pluto moves in the distance for most of its rotational period; it cannot be observed near the ecliptic.

IV.
THE ZODIAC

The zodiac[46] is a band or belt on either side of the ecliptic extending approximately 8°-9° northward and southward, with a breadth of 16°-18° in total. As the Sun, Moon, and other planets go on with their motions, they appear in various constellations in this belt at various times. Only Pluto can't be seen around the ecliptic because it has a 17.15° inclination to the plane of the ecliptic.

Only 12 of the 88 constellations identified on the celestial sphere are located over this virtual belt. At this point, the constellation Ophiuchus, which occasionally appears in the news and is referred to as the 13th sign[47] of the zodiac, bears mentioning. This constellation occupies a part of the zodiac above the constellation Scorpio. Astronomically, the Sun is in the constellation Ophiuchus from November 30 to December 17.

It was the Babylonians who removed the constellation Ophiuchus from the zodiac belt and left us with the 12-sign zodiac, as well as the current names of the signs. This formation of the zodiac belt originates from cuneiform tablets tablets called MUL.APIN, which date back to roughly 700 BC, predating the Babylonians. However, these tablets feature the astronomic sizes of the signs.

46 Zodiac comes from the Latin word zōdiacus, which means "circle of animals."

47 The Babylonians observed that the zodiac belt consisted of 13 constellations.

Image 14. The zodiac

Each constellation occupies a different amount of space in the zodiac.[48] In western astrology, which is based on the tropical zodiac, it is not the constellations that we qualify as the signs of the zodiac. In order to get a better understanding of this, we must first define the tropical zodiac.

The Tropical Zodiac

The tropical zodiac is concerned with the seasons, which is why it is also known as the "the zodiac of the seasons." It is defined according to Hipparchus's discovery of the shifting of the vernal point in the 2nd century BC and the star catalogue that Ptolemy devised based on his observations. With that said, there is a lot of evidence suggesting that the Babylonians had this knowledge before Hipparchus. The tropical zodiac rose to popularity in the west after the 6th century AD.

48 Astronomically, constellation Aries occupies 24°; constellation Taurus, 36°; constellation Gemini, 28°; constellation Cancer, 21°; constellation Leo, 35°; constellation Virgo, 46°; constellation Libra, 18°; constellation Scorpio, 31°; constellation Sagittarius, 30°; constellation Capricorn, 28°; constellation Aquarius, 25°; and constellation Pisces, 38° in the zodiac.

The tropical zodiac is measured by the annual passage of the Sun through the vernal equinox. This zodiac begins at 0° of Aries and is regulated according to the seasons by being divided into 12 equal sections, beginning at 0° of Aries, taking 30° per each section. In order to understand the relationship between the tropical zodiac and the seasons, let's take a look at the seasons first.

THE SEASONS

The primary reasons for the formation of the seasons are the Earth's revolution around the Sun and the Earth's equator's 23.5° incline to the orbital plane. Seasons are parts of the year with distinct astronomical and climatic characteristics. There are four of them: spring, summer, autumn, and winter.

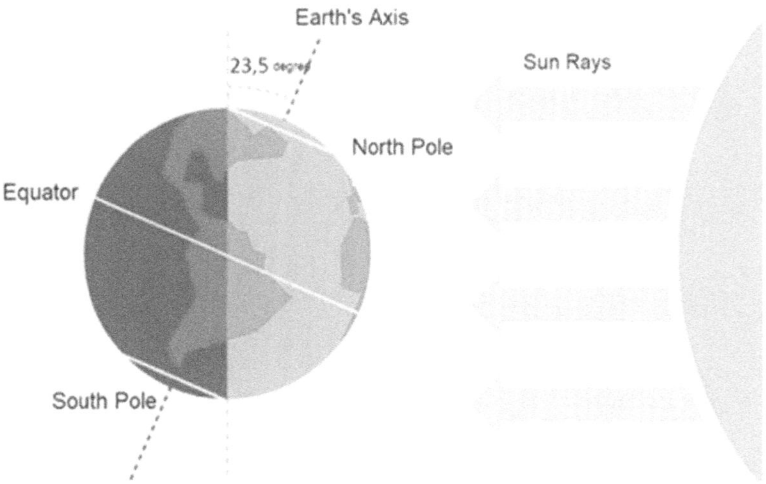

Image 15. The axial tilt of the earth

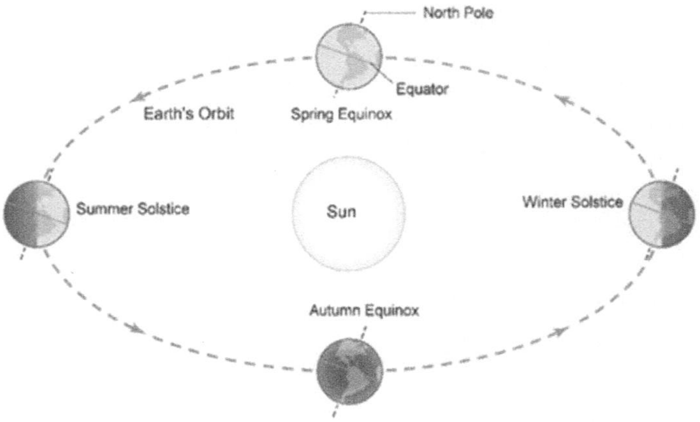

Image 16. The seasons

The Earth has two different motions: rotating around itself and revolving around the Sun. The Earth revolves in its ecliptic orbit with the Sun in its focus. The point where the Earth is closest to the Sun is called perihelion (perihel), and the point where it is farthest from the Sun is called aphelion (aphel). The Earth is at perihelion around January 2nd and at aphelion around July 4th.

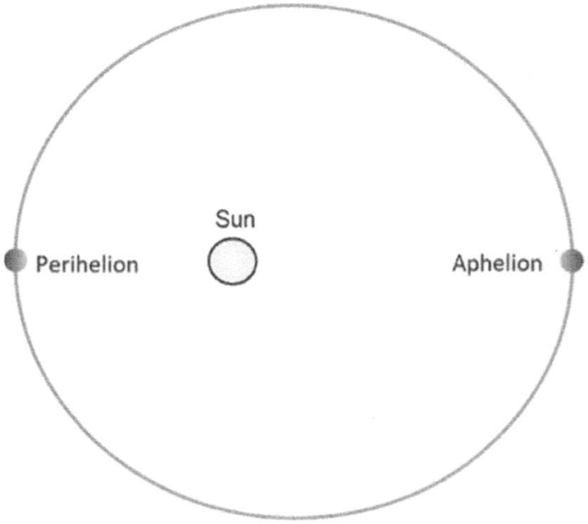

Image 17. Perihelionaphelion

The fundamental reason why the northern and southern hemispheres warm up and cool down at different times is the angle at which the Sun's rays strike the surface of the Earth. During the months when the Earth is farthest from the Sun, which are June and July, it is summer in the northern hemisphere. This is because the angle at which sun-beams strike the ground increases in June in the northern hemisphere. When it's summer in the northern hemisphere, it is winter in the southern hemisphere. Likewise, when it is autumn in one hemisphere, it is spring in the other hemisphere.

What Are Equinox and Solstice Points?

The beginnings of the seasons, which are caused by the incline of the Earth's rotational axis and its yearly motion, have special names. The beginnings of the four seasons—spring, summer, autumn, and winter—coincide with equinox and solstice points. Equinoxes occur on or around March 21st and September 22nd. Solstices occur on December 21st and June 21st.

There is a 23°27′—i.e., 23,5°—angle between the plane of the ecliptic and the plane of the celestial equator. The intersection points of the two planes are called equinoxes. These are the vernal and autumnal equinoxes. The vernal equinox is the point where the Sun crosses the celestial equator as it moves northward. Inversely, the autumnal equinox is the point where the Sun crosses the celestial equator as it moves southward. The vernal equinox occurs when the Sun is at 0° Aries and the autumnal equinox occurs when the Sun is at 0° Libra.

The vernal equinox occurs approximately on March 21 every year. On this date, both hemispheres receive the same amount of sunlight. Spring begins in the northern hemisphere and autumn begins in the southern hemisphere. The autumnal equinox occurs approximately on September 22 every year. It marks the beginning of autumn in the southern hemisphere and of spring in the northern hemisphere. On equinox dates, the rays of the Sun hit the equator perpendicularly, and day and night are of equal duration throughout the Earth.

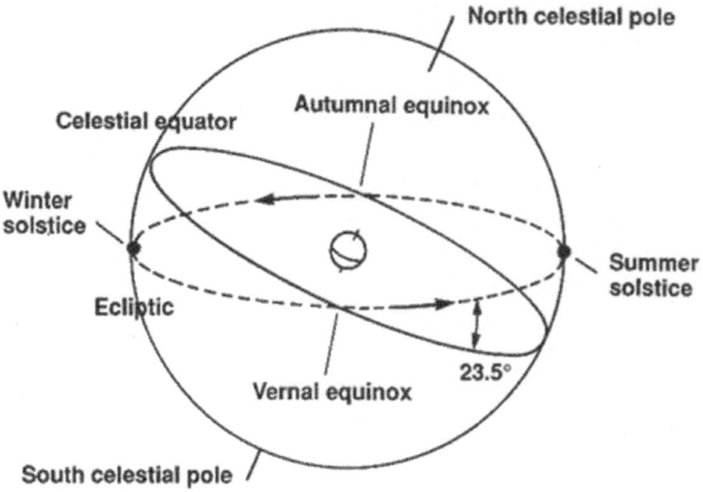

Figure 6. Equinox and solstice points

The points where the Sun reaches its maximum distance from the south and north of the celestial equator are called "solstices." Every year, around December 21, the Sun arrives at the winter solstice point, and those located in the northern hemisphere experience the longest night, or the shortest day, of the year. The winter solstice occurs when the Sun is at 0° Capricorn. On June 21, the Sun rises and sets at its northernmost point. This date marks the longest day and the shortest night of the year. Solstices are the dates on which days begin to get longer or shorter. The summer solstice point begins at 0° Cancer. After June 21, the Sun's rising and setting points shift southward, the days get shorter, and the nights get longer. On December 21, the Sun rises and sets at its southernmost point.

These points that initiate the seasons also mark the beginnings of the four cardinal signs of astrology—Aries, Cancer, Libra, and Capricorn. One of the arguments devised against astrology is that the vernal equinox is currently in the constellation Pisces. This is astronomically correct, but even though the Sun is in the constellation Pisces at the vernal equinox in the tropical zodiac, it is still considered to be in Aries because the tropical zodiac is based not on the constellations but on the beginnings of the seasons—it doesn't take into consideration the shift in star positions. Now let's have a look at the sidereal zodiac, which does take this shift into consideration.

The Sidereal Zodiac

The sidereal zodiac takes a fixed star as reference. Like the tropical zodiac, the sidereal zodiac consists of 12 signs, 30° each. It differs from the tropical zodiac in that it takes precession into consideration.

What Is Precession?

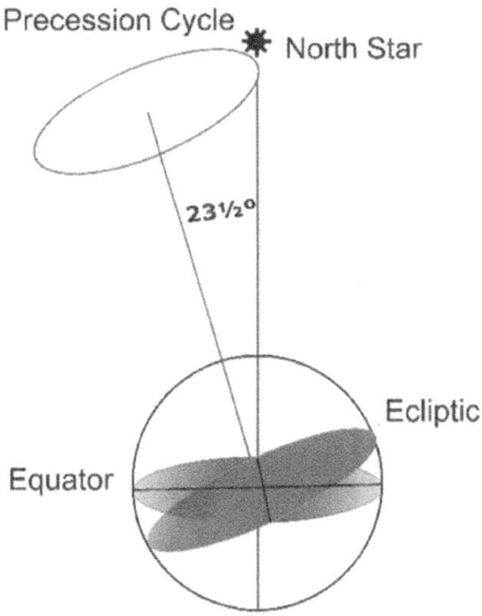

Figure 7. Precession

Because of the gravitational pulls of the Sun and the Moon, the Earth's rotational axis has an incline of 23°27′ with respect to its orbital plane. Again, because of gravitational pulls, the Earth's rotational axis exhibits a slow yet constant shift. We call this shift precession. The rotational axis moves by 360° every 25,772 years, like a spinning top, and it changes its direction. The precessional motion of the Earth causes the equinox points to shift. Every 72 years, the equinox points shift westward on the ecliptic by 1° with respect to the stars in the sky. In other words, in 2160 years, the zodiac will have shifted by 30°. The sidereal zodiac takes this shift into consideration.

Although the sidereal zodiac has been documented since the 5th century BC, precession is thought to have been discovered later, in the 2nd century BC. We know that the Babylonians used the sidereal zodiac, despite knowing about the precession movement.

In the sidereal cycle, the vernal point shifts backward. The point of Aries remained in the constellation Aries from 1000 BC to 420 AD. Afterward, this point shifted to the constellation Pisces because of the Earth's precession.

The sidereal zodiac can be defined under three headings:

1. The Babylonian Zodiac

The foundations of the Babylonian zodiac date back to the early 5th century BC. This zodiac is defined along the Aldebaran-Antares axis. Aldebaran and Antares are stars at the center of the constellations Taurus and Scorpio, respectively. The Babylonian zodiac forms the basis of the zodiac that we use today, which consists of 12 equal pieces of 30° each.

2. The Indian Zodiac

The Indian zodiac uses the sidereal zodiac, which takes the stars as reference. This point is the fixed star Spica (Alpha Virginis).

3. The Astronomical Zodiac

The astronomical zodiac defines the signs by their true widths on the zodiac. The areas of the constellations are designated in accordance with the definitions accepted by the IAU (International Astronomical Union) in 1930. Some sidereal astrologers evaluate the signs based on their true sizes instead of considering their sizes equal.

V.
COORDINATE SYSTEMS

Geographical Coordinate System

We use geographical coordinates to define a point on Earth. These coordinates are expressed as longitude and latitude.

Latitude is the angular distance of a point on Earth from the equator. To find the latitude, we take the equator as 0°. Values above the equator are positive, while those below the equator are negative: latitude is measured between 0° and +90° north of the equator, and between 0° and –90° south of the equator. Figure 9 shows the geographical coordinates for a point A on Earth. Let's assume that point A is the city of İzmir. In this case, we can say that the latitude of İzmir is 38°25N (N stands for north) or +38°25. (You can find your latitude on Earth by using the pole star[49]. The angle between the pole star and the plane of the horizon of your location equals your latitude.)

Longitude is the angular distance of a point on Earth from the Greenwich meridian[50]. Longitude is measured between 0° and –180° eastward along the equator and between 0° and +180° westward along the equator. For instance, the longitude of İzmir is 27°09E (E stands for east) or +27°09. Geographical coordinates are used in calculating the natal chart.

---

49 You can learn how to find the pole star by reading the chapter titled "Celestial Observation" in this book.

50 Meridians are the large circular arcs that connect the Earth's north and south poles and run perpendicular to the equator. The Greenwich meridian is the meridian that runs through the Greenwich Observatory, and its longitude is 0°. It is also known as the "Prime Meridian."

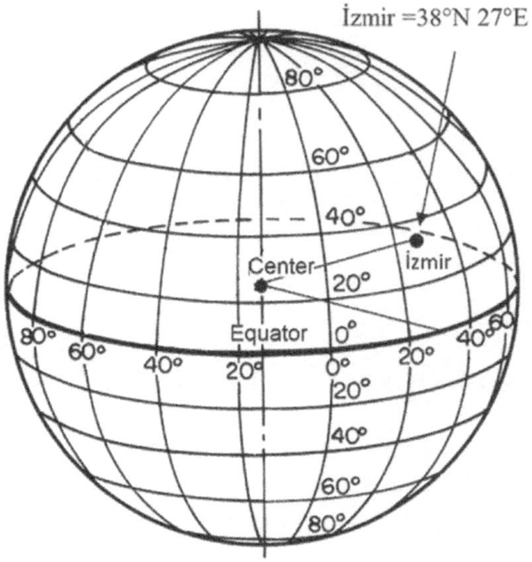

Figure 8. The latitude and longitude of Izmir, shown on the globe

Celestial Coordinates

Although geographical coordinates can be used to define a point on Earth, they cannot be used to define a celestial body. The star maps made from antiquity to the present are, in fact, the results of an effort to determine and define the locations of celestial bodies. These maps were created in a variety of coordinate systems. Today, it is important to know which coordinate system one is discussing in order to create a common celestial language and interpret the marks inherited from the past. Whether we are observing the sky with a simple telescope or a state-of-the-art satellite, we need common terms when recording our findings. For this, we use the celestial sphere as a reference.

All the stars and planets that we see in the sky are at various distances from us. However, we cannot perceive this distance in the sky; to us, all celestial bodies appear to be pinned to a sphere. This virtual sphere on which all stars or planets are thought to be situated, with a massive radius and the Earth at its center, is called the celestial sphere. You could think that calling it a "virtual sphere" makes it sound trivial—on the contrary, it is a rather significant notion. Calculations performed on this sphere, for example, can aid in determining the latitude and time of a solar eclipse that will occur 60 years later. Because astrology is concerned with the motions of celestial bodies observable from Earth, the celestial sphere and celestial coordinate systems take on more significance. So, let's begin by getting to know the celestial sphere better.

Our Earth rotates on its own axis from west to east. The axis of the celestial sphere is the same as the Earth's rotational axis, and all celestial bodies revolve around it. Consider how the stars and planets appear to move from east to west when viewed from Earth. The same motion occurs on the celestial sphere, with all celestial bodies rising from the east and setting in the west.

The great circle perpendicular to the rotational axis of the celestial sphere is called the "celestial equator." The celestial equator is the Earth's equator, widened and projected onto the celestial sphere. It splits the celestial sphere into two equal halves—north and south. The points where the axis of rotation pierces the Earth are called terrestrial poles, and the points where the axis of rotation pierces the celestial sphere are called celestial poles.

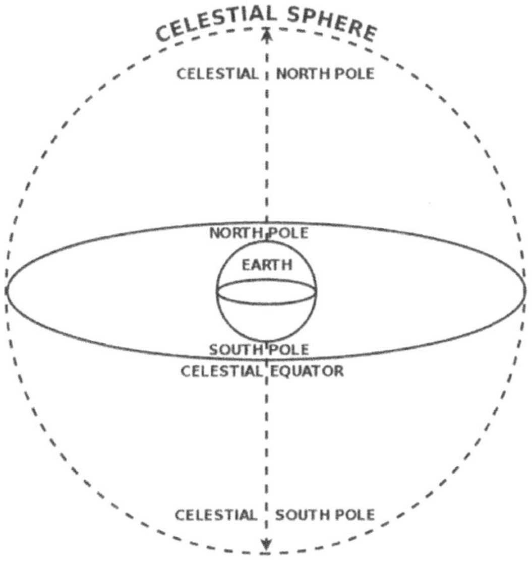

Figure 9. The celestial sphere

The pole visible from the northern hemisphere is known as the celestial north pole, and the pole visible from the southern hemisphere is known as the celestial south pole.

In order to specify the location of a certain celestial body on the celestial sphere, we use coordinate systems based on the ecliptic, the equator, or the circle of the horizon. Now, let's examine these three coordinate systems in further detail.

a. The Ecliptic Coordinate System

The Earth completes its revolution around the Sun in one year. When viewed from the Earth, the Sun appears to revolve around the Earth. The great circle along which the Sun travels over the course of a year is called the "ecliptic." The ecliptic was first defined around 2000 BC. It gets its name from the fact that lunar and solar eclipses occur near the plane of the ecliptic, which was discovered a long time ago.

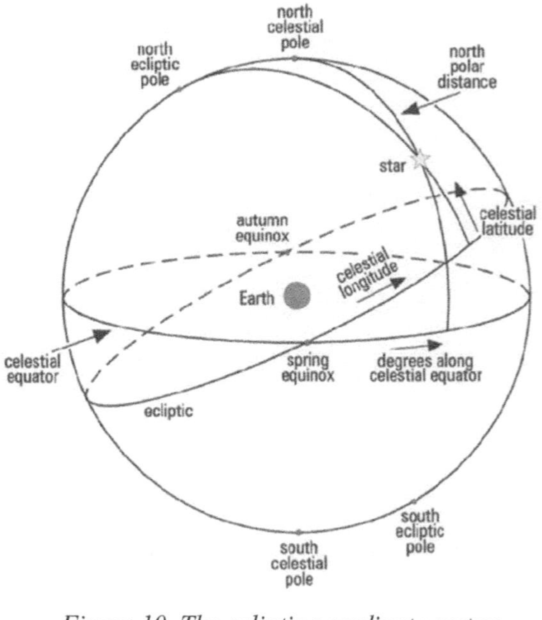

Figure 10. The ecliptic coordinate system

The ecliptic coordinate system is based on the plane of the ecliptic and is independent of observation location and time. Astronomers of the antique era utilized this coordinate system to establish the positions of the Sun, Moon, planets, and other celestial bodies on the ecliptic.

The ecliptic system uses two coordinates: celestial longitude and celestial latitude. Celestial latitude is the angular distance of any celestial body from the plane of the ecliptic. This distance is measured in degrees north and south from the ecliptic poles, perpendicular to the plane of the ecliptic. It is measured between 0° (the plane of the ecliptic) and +90° (the ecliptic north pole), or between 0° and −90° (the ecliptic south pole). Celestial longitude is the distance between a circle passing through the pole of the ecliptic and a star and another circle passing through, again, the pole of the ecliptic and the vernal point (0° Aries). It is measured between 0° (the vernal equinox) and 360° in positive angles. The degrees and minutes of the Sun, Moon, and all the other planets given in ephemerides[51] correspond to celestial longitude.

CHART POINTS

Point	Longitude	Travel	Latitude	Rt.Asc.	Decl.	Azi(0°N)	Alti.
☽	04°♋54'24"	+12°26'	+04°53'	095°33'	+28°14'	334°47'	-06°21'
☉	16°♌26'12"	+57'32"	+00°00'	138°53'	+15°54'	293°45'	+01°43'
☿	22°♌28'16"	+01°57'	+01°44'	145°24'	+15°39'	288°35'	+05°18'
♀	01°♋30'09"	+49'56"	-04°05'	091°35'	+19°20'	335°45'	-15°54'
♂	09°♈19'20"	+14'01"	-04°52'	010°28'	-00°47'	062°57'	-20°49'
♃	02°♊44'50"	+08'08"	-00°56'	060°53'	+19°46'	005°54'	-18°31'
♄	26°♐17'57" ℞	-02'01"	+01°02'	265°59'	-22°21'	161°45'	+14°15'
♅	27°♐21'18" ℞	-01'18"	-00°12'	267°06'	-23°37'	161°00'	+12°47'
♆	07°♑50'51" ℞	-01'10"	+00°58'	278°28'	-22°13'	150°19'	+11°13'
♇	09°♏52'09"	+00'40"	+15°56'	222°26'	+00°23'	210°15'	+34°53'
⚷	04°♋15'23"	+04'36"	-05°46'	094°26'	+17°36'	332°31'	-16°48'
☊	15°♓28'35" ℞	-03'10"	+00°00'	346°37'	-05°43'	085°54'	-10°33'
As	11°♒07'59"	+606°02'	+00°00'	313°35'	-17°26'	118°46'	-00°00'
Mc	08°♐34'28"	+339°30'	+00°00'	246°50'	-21°44'	180°00'	+16°45'

Image 18. Celestial longitudes, as shown on the Solar Fire program

b. The Equatorial Coordinate System

The equatorial coordinate system is used to express the positions of celestial bodies in space and relative to one another. It is similar to the geographical coordinate system. When the latitudes and longitudes on Earth are projected onto the celestial sphere, latitude corresponds to "declination" and longitude corresponds to "right ascension."

51 Ephemerides are sky journals; they contain the positions of the celestial bodies.

Declination is the angular distance of a celestial body from the celestial equator. It is measured between 0° and +90° north of the equator and 0° and +90° south of the equator. Right ascension is the angular distance of the great circle passing through the celestial body from the vernal equinox, or 0° Aries. It is measured along the equator, beginning at 0° Aries and ranging from 0 to 24 hours in the positive direction. This is because the Earth takes 24 hours to rotate around its own axis. The Earth is like a giant clock that rotates around itself, and each hour of time changes the right ascension by an hour.

Right ascension and declination values have nothing to do with the location of the observer. They are a common coordinate system that can be used by anyone on Earth. So long as the direction of the Earth's axis of rotation does not change, the equatorial coordinates of celestial bodies almost remain the same. *Almost,* because the precession movement in the Earth's axis of rotation causes some changes, albeit minor. Thus, in star catalogues and atlases, equatorial coordinates are listed along with the year to which they belong.

What Is An Hour Angle?

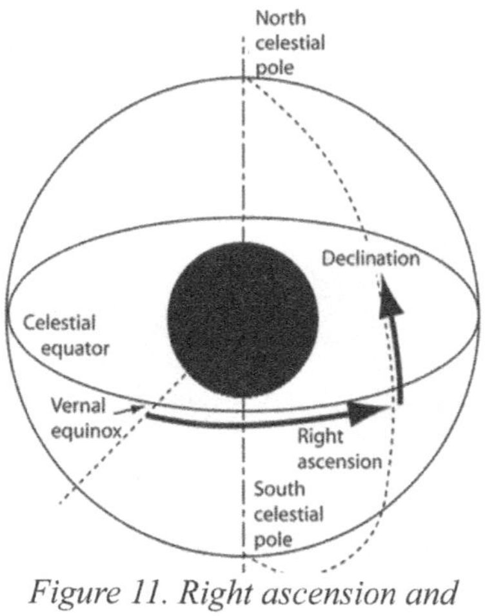

Figure 11. Right ascension and declination

The hour angle is used in the equatorial coordinate system, which takes time and the location of observation into account. The hour angle is the angular distance measured westward between the vertical circle (the hour circle) passing from the star to the observer's meridian[52]. It ranges between 0 and 24 hours. The other coordinate is declination, which is measured as described in the equatorial coordinate system.

52 The observer's meridian is the great circle passing through the observer's zenith, the celestial north pole, and the celestial south pole. Throughout its diurnal motion, the Sun reaches its highest point at noon.

Parallels and Contra-Parallels

In astrology, parallel and contra-parallel aspects are expressed according to declination. For example, suppose a planet's declination is +18° and another planet's declination is also +18°. In this case, we call these two planets "parallel." If a planet's declination is +18° and another planet's declination is −18°, then these two planets are "contra-parallel." You can access parallel and contra-parallel aspects from the reporting menus of astrology programs.

CHART POINTS

Point	Longitude	Travel	Latitude	Rt.Asc.	Decl.	Azi(0°N)	Alti.
☽	04°♋54'24"	+12°26'	+04°53'	095°33'	+28°14'	334°47'	-06°21'
☉	16°♌26'12"	+57'32"	+00°00'	138°53'	+15°54'	293°45'	+01°43'
☿	22°♌28'16"	+01°57'	+01°44'	145°24'	+15°39'	288°35'	+05°18'
♀	01°♋30'09"	+49'56"	-04°05'	091°35'	+19°20'	335°45'	-15°54'
♂	09°♈19'20"	+14'01"	-04°52'	010°28'	-00°47'	062°57'	-20°49'
♃	02°♊44'50"	+08'08"	-00°56'	060°53'	+19°46'	005°54'	-18°31'
♄	26°♐17'57" ℞	-02'01"	+01°02'	265°59'	-22°21'	161°45'	+14°15'
♅	27°♐21'18" ℞	-01'18"	-00°12'	267°06'	-23°37'	161°00'	+12°47'
♆	07°♑50'51" ℞	-01'10"	+00°58'	278°28'	-22°13'	150°19'	+11°13'
♇	09°♏52'09"	+00'40"	+15°56'	222°26'	+00°23'	210°15'	+34°53'
⚷	04°♋15'23"	+04'36"	-05°46'	094°26'	+17°36'	332°31'	-16°48'
☊	15°♓28'35" ℞	-03'10"	+00°00'	346°37'	-05°43'	085°54'	-10°33'
As	11°♒07'59"	+606°02'	+00°00'	313°35'	-17°26'	118°46'	-00°00'
Mc	08°♐34'28"	+339°30'	+00°00'	246°50'	-21°44'	180°00'	+16°45'

Image 19. Parallels and contra-parallels

c. The Horizontal Coordinate System

Coordinate systems defined on the horizon circle are called horizontal coordinate systems. Using this system, the observer can determine the location of a celestial body at a specific time and location.

The "horizontal plane" refers to the location of an observer on Earth. Let's say you're sailing on a ship in the middle of the ocean. If there is no land in sight when you look around, the horizon is where the water meets the celestial sphere. Draw a line from your point that is tangential to the terrestrial sphere; the plane you obtain is the plane of the horizon. The points where this plane intersects the celestial sphere constitute the horizon circle. If a celestial body moves downward from the horizon circle, it is setting; if it moves upward from the horizon circle, it is rising. Any celestial body below this plane will be invisible to an observer. The Sun, for example, rises along the eastern horizon in the morning and rises over the horizon throughout the day. It reaches its maximum altitude when it passes the observer's meridian. After that point, it begins to descend, eventually setting at a point on the western horizon directly across from the point on the eastern horizon where it rose.

Now, let's imagine a perpendicular line erected from the observation point. This line intersects the celestial sphere at two points: the zenith and the nadir. The zenith is directly overhead; you can see it when you lift your head straight up during observation. The nadir is located in the part of the celestial sphere that we cannot see. Since there are countless observation points on Earth, there are also countless zeniths and nadirs.

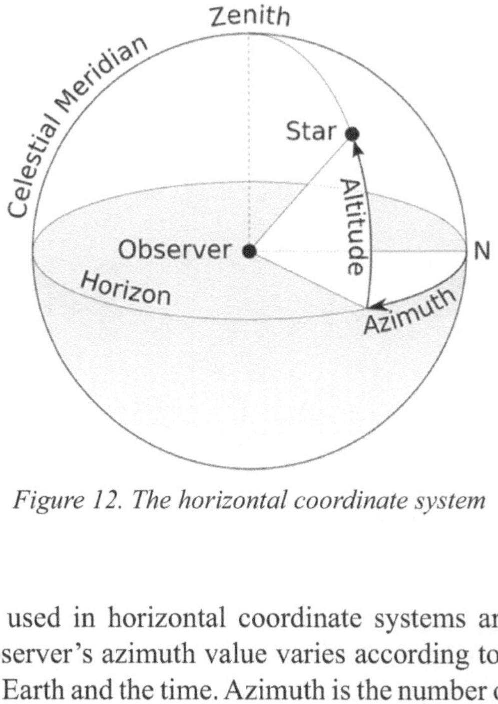

Figure 12. The horizontal coordinate system

The angles used in horizontal coordinate systems are azimuth and altitude. An observer's azimuth value varies according to the observer's location on the Earth and the time. Azimuth is the number of degrees from the north point to the great circle passing through the star and the zenith, measured clockwise (eastward) along the horizon circle. It is measured between 0° and 360° in the negative direction from the south point. Some measure the azimuth starting from the south point, although it is, again, measured in a clockwise direction. Altitude is the vertical angular distance of a star from the plane of horizon. It is measured between 0° and 90° upward from the horizon, toward the zenith, and between 0° and −90° downward from the horizon.

VI.
HOUSE SYSTEMS

Houses are components of the natal chart that make the chart more personal and unique. We need the exact time and date of birth in order to calculate the house placements. The houses are created by symbolically dividing the sky into 12 sections. 6 of these houses are below the horizon, while the other 6 are above it. Each house represents a different aspect of life. Houses are numbered in a counter-clockwise direction. The lines dividing the houses are called "cusps." Each house is determined by the longitude angle of the sign that intersects the house.

Although we don't know if houses were used in the antique period, we do know that the Babylonians attached importance to the rising and setting of the planets. The Egyptians not only mythologized the Sun's journey throughout the day, but they also gave it meaning. Other civilizations recognized the significance of each direction as well. Whether a planet was in the north, south, east, or west meant something. In fact, the inter-cardinal directions, like the cardinal directions, were associated with the gods and used for prophesying. This entire historical process has paved the way for the formation of house systems.

Astrologists attach importance to four points in the natal chart: the ascendant (ASC), the midheaven (MC, *medium coeli*), the nadir (IC, *imum coeli*), and the descendant (DSC). Each degree of each sign rises in the east and then sets. The eastern horizon is known as ASC; the western

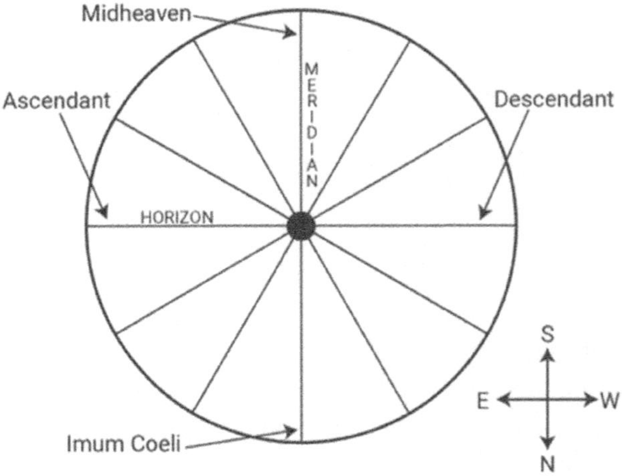

Figure 13. ASC-MC-DSC-IC placements

horizon, DSC; the middle of the eastern and western horizons, MC; and the lowermost point, IC. The ascendant is the point in the east where the plane of the horizon intersects the plane of the ecliptic. The descendant is the point in the west where the plane of the horizon intersects the plane of the ecliptic. The MC is the point where the ecliptic and the meridian intersect above the horizon, and the IC is the point where the ecliptic and the meridian intersect below the horizon.

The signs rise uniformly at the equator. Due to the obliquity of the ecliptic, some signs rise over the horizon faster than others. At northern latitudes, signs from Cancer to Sagittarius rise quickly (long ascension), and signs from Capricorn to Gemini rise slowly (short ascension). At southern latitudes, the exact opposite of this is in question.

House Systems

Until the 15th century, astrologists divided the sky into 12 sections using several house systems. Which house system to choose is a much-debated topic among astrologers. A natal chart created according to one house system may place a planet in a certain house, whereas a natal chart created according to another house system may place the same planet in another house. An astrologer may experiment with various house systems and settle on a specific one based on his or her experiences. Different systems have different benefits of their own. Each house system can help in interpreting a natal chart from a different perspective. For instance, the ecliptic-based systems provide more insight into the self as they are based on the Sun's journey through-out the year. The most important of the 25 house systems discovered can be divided into three categories:

1. The Ecliptic Systems
2. The Space Systems
3. The Time Systems

1. The Ecliptic Systems:

These systems are created by dividing the ecliptic. We will be looking into the Equal House and the Porphyry systems, as well as the Whole Sign system, since it also entails dividing the ecliptic.

a. **Whole Sign:** In the Whole Sign house system, which is among the oldest of its kind, the order of the zodiac signs aligns with the houses. The first house begins with 0° of the relevant sign. So, the 1st house completely encapsulates the sign placement of ASC. The 0° of the sign that follows the ASC serves as the cusp of the 2nd house. The 3rd and 4th houses proceed in the same manner.

The whole sign system seems hard to embrace for modern astrologists, as the ASC and MC do not coincide with the 1st and 10th house cusps.

b. **Equal House:** This is among the oldest house systems, dating back to the 3000s BC. Although it wasn't used for a while, Alan Leo popularized it roughly thirty years ago. It is still in use today.

This system doesn't require any complicated mathematics. The ecliptic is divided into 12 equal sections. All house cusps are determined by starting at the degree of the ASC and taking 30° for each house. For instance, if your ASC is 20° Cancer, then the 2nd cusp will be 20° Leo, the 3rd cusp will be 20° Virgo, the 4th cusp will be 20° Libra, and so on.

In the equal house system, the ASC is taken as the cusp of the 1st house. The MC-IC axis, often known as the 'parental axis' in modern astrology, does not coincide with the cusps of the 10th and 4th houses, but astrologists who evaluate the natal chart from a psychological point of view use this system to interpret human behaviors and characters.

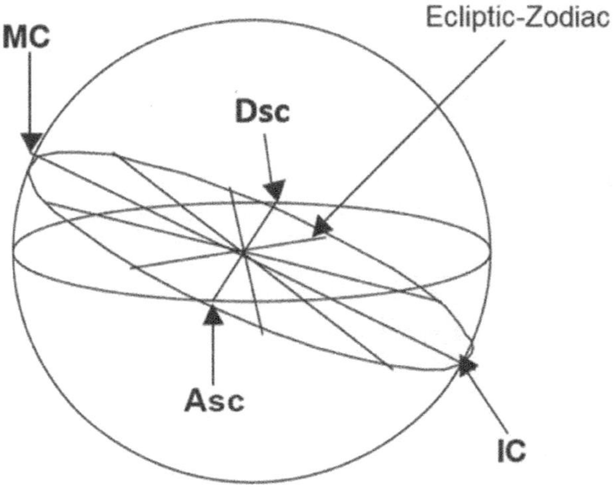

Figure 14. The equal house system

c. **Porphyry:** This is one of the quadrant systems. Although this system was devised by Porphyry of Tyre in the 3rd century BC, it was first defined in Vettius Valens's The Anthology in the 2nd century.

In this system, which is a different type of equal house system, the quadrants are divided into equal pieces. In short, this system places the MC on the 10th cusp and the ASC on the 1st cusp. The eastern quadrants on the ecliptic are then trisected, from ASC to MC and from ASC to IC, to calculate the cusp degrees of the 11th and 12th houses, as well as the 2nd and 3rd houses.

Let's clarify this through an example:

Say MC is in 10° Gemini, and ASC is in 13° Virgo. First, we will calculate the difference between MC and ASC and divide it by 3:

MC in 10° Gemini=70° (This value corresponds to 70° when expressed between 0°-360°)

ASC in 13° Virgo=163° (This value corresponds to 163° when expressed between 0°-360°)

The difference 163−70=93°/3=31°

Adding 31° to MC gives us the cusp of the 11. house.

70°+31°=101°, which corresponds to 11° Cancer. So, the cusp of the 11. house is 11° Cancer.

Adding 31° (difference of ASC and MC, divided by 3) to this cusp gives us the cusp of the 12. house.

Now, let's do the same operations on the nocturnal semi-arc between the IC and the ASC.

IC is 10° Sagittarius=190°

ASC is 13° Virgo=163°

The difference 190−163=26/3=8,6°

Then we calculate the house cusps based on the one-third degrees that we have obtained:

The cusp of the 2. house is 12° Libra and the cusp of the 3. house is 11° Scorpio.

The Porphyry system has fallen out of favor in recent years. Although ecliptic-based systems provide a true projection of the sky, their divisions cause problems.

2. The Space Systems:

The space systems are based on dividing circles other than the ecliptic (the celestial equator, the horizon, and the prime vertical[53]) and using the points where they intersect with the ecliptic as the house cusps. For this, arcs extending from the north to the south of the horizon are used. The main space-based house systems are the Campanus, Morinus, and Regiomontanus.

a. **Campanus:** The Campanus system was defined by Johannes Campanus, who lived in the 13th century. The cusps are determined

53 The prime meridian is the great circle passing through the zenith and nadir and intersecting the horizon circle at the west and the east.

by dividing the prime vertical into equal sections of 30° each, and then projecting these onto the ecliptic.

This system stands out with its emphasis othe planet's position in relation to the horizon and meridian at the place of birth. Its disadvantages are that, in theory, it undermines the role of the ecliptic and, in practice, it falls short in high latitudes.

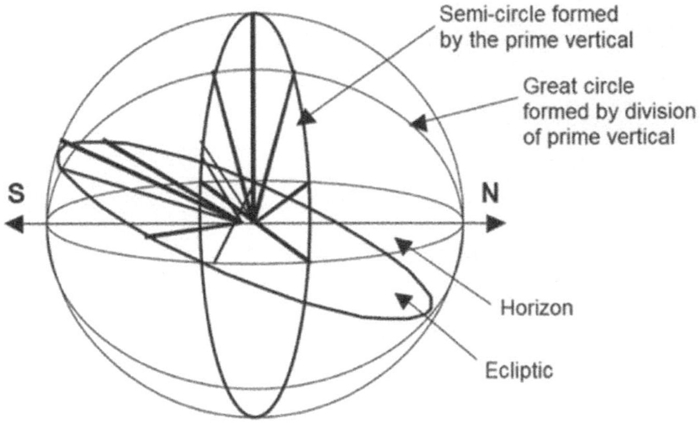

Figure 15. The Campanus house system

b. **Regiomontanus:** The Regiomontanus house system was devised by Johannes Müller, a mathematician who lived in the 15th century and was also known as Regiomontanus. It might be thought of as an enhanced version of the Campanus system. In this system, the celestial equator is divided into equal sections of 30° and projected onto the ecliptic.

In this system, ASC and MC intersect the 1st house and the 10th house, respectively. It emphasizes the Earth's daily rotation around itself. It became quite popular in the 17th century, and was used by many astrologists at the time, including William Lilly. This system is still used today.

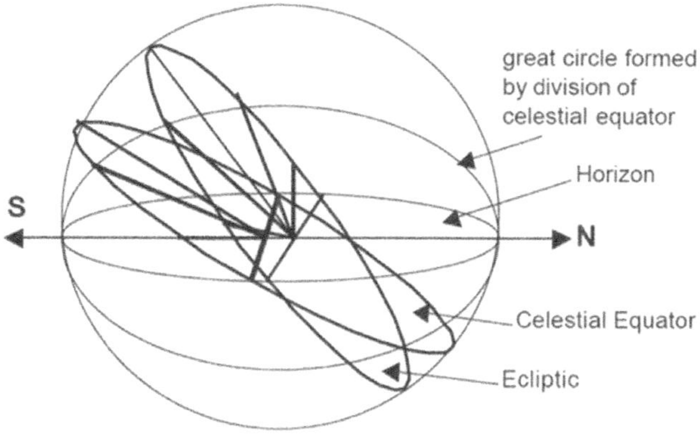

Figure 16. The Regiomontanus house system

c. **Morinus:** This system was devised by the French astrologist and mathematician Jean-Baptiste Morinus. In this system, which can also be used at high latitudes, the celestial equator is divided into equal sections and projected onto the ecliptic. It differs from Regiomontanus, which divides the celestial equator into equal sections and projects them onto the ecliptic, in that this system projects these equal sections onto the ecliptic via longitude circles. Thus, the ASC does not coincide with the cusp of the 1st house, which is why this system has never really gained popularity.

3. The Time Systems:

The time systems are based on the time it takes for any degree to move along the axes. For instance, the time it takes for a degree to rise from the ASC to the MC is calculated. This time is then divided into equal sections to obtain various house systems. The most popular among these are the Placidus, Koch, and Alcabitius systems.

In these house systems, ASC marks the cusp of the 1st house; DSC, the 7th house; MC, the 10th house; and IC, the 4th house. These houses pose a problem at high latitudes.

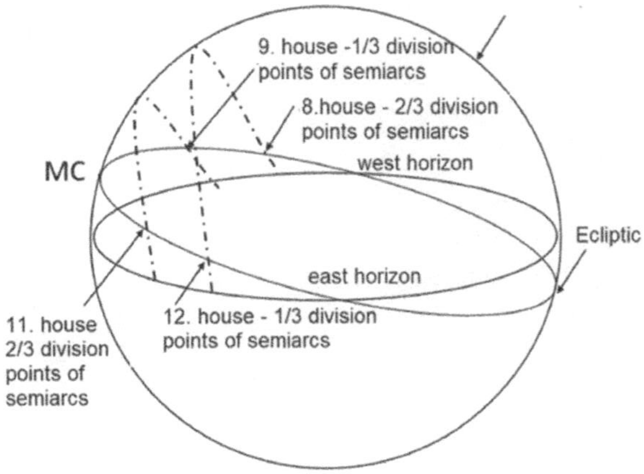

Figure 17. The Placidus house system

a. **Placidus:** This system was popularized in the 17th century by the Italian monk Placidus de Titis. It is among the most widely used house systems in western astrology. Placidian tables are common and easily accessible, which is why this system is so popular.

Unlike the other house systems, the Placidian system isn't based on equally dividing the circles. First, the time it takes for the ASC (ascending) to rise to the MC (midheaven) is determined. This time is then trisected to find the cusps of the 12th and 11th houses. Likewise, the time it takes for a degree to rise from the IC to the ASC is trisected to determine the cusps of the 2nd and 3rd houses. The cusps of the 8th, 9th, 5th, and 6th houses are easy to calculate, as they are right across from these houses.

Although the calculation of cusps appears to be simple in logic, the calculations that involve trigonometry are rather complex. The MC always coincides with the cusp of the 10th house, but some houses may intersect two or three signs at the same time. This irregularity is caused by the fact that the division times in the semi-diurnal arc and the semi-nocturnal arc are different.

This system falls short in latitudes higher than 66.5°, because certain degrees of the ecliptic are known as "circumpolar," which means the stars and planets never rise here, and there is no ASC. Thus, the time intervals between the axes cannot be calculated. The Placidian house system is inconvenient in this regard. Yet it still continues to be among the most widely used quadrant house systems in the present day.

b. Koch: Also known as the "birthplace system," this system rose to popularity because of the tables published by its creator, Walter Koch, in 1971. This system was created based on the ascendant and the birthplace. It requires quite complex trigonometric calculations. Like the Placidus system, the Koch system is also inconvenient in high latitudes. Nonetheless, it is a popular house system.

c. Alcabitius: This system gets its name from Alcabitius, an Arab astrologist who lived in the 12th century. It was defined by Rhetorius, a Hellenistic astrologist that has lived around 500s BC, before Placidus's time. In the earlier periods, it was the most widely used quadrant system.

In this system, the time interval between the ASC and the MC is trisected. These equal sections are then projected onto the ecliptic. This house system, which isn't popular in the present day, produces more accurate results at higher latitudes compared to the Placidus system.

d. Meridian (Zariel): Also known as the "axial rotation house system," this system was devised by David Cope (under the pen name "Zariel") in the 1900s. The celestial equator is divided into 12 equal sections of 30°, starting from Aries 0°. These sections are projected onto the ecliptic by hour circles. The cusp of the 10th house intersects the MC, but the ascendant is equal to the equatorial ascendant. This system does not pose a problem in high latitudes, but it receives criticism as it does not take the horizon of the birthplace into consideration. It is mostly used by Uranian astrologists.

VII.
PLANETS IN RETROGRADE

All planets revolve in a counter-clockwise direction around the Sun, from west to east. At times, however, the planets appear to be moving backwards. What really happens then is, in fact, similar to how a driver who stops at a red light thinks his own car is moving backwards when he sees the car next to him moving forward. Retrograde motions are caused by the different speeds at which the Earth and the other planets revolve around the Sun. Thus, when looking from Earth, we first see the planet slow down, then pause and move backwards.

The alignments formed during the retrograde motions of the inferior planets and superior planets differ. The inferior planets are Mercury and Venus. These planets are in closer orbital proximity to the Sun, and they move within a narrower space. The superior planets are Mars, Jupiter, Saturn, Uranus, Neptune, and Pluto. These planets are further from the Sun than the Earth. The main difference between the retrograde motions of the inferior and superior planets is their position relative to the Sun, and their only common point is that they are all closest to the Earth when in retrograde.

Retrograde Motion Of The Inferior Planets

Mercury and Venus make two kinds of conjunctions with the Sun. One of these is superior conjunction, which is when the Sun comes between the inferior planet and the Earth, and the inferior planet appears to move forward. The other is inferior conjunction, which is when the Earth, the inferior planet, and the Sun all align in that order. In this case, the planet appears to make a retrograde motion. Although Venus and Mercury have different retrograde cycles, the system operates in precisely the same way.

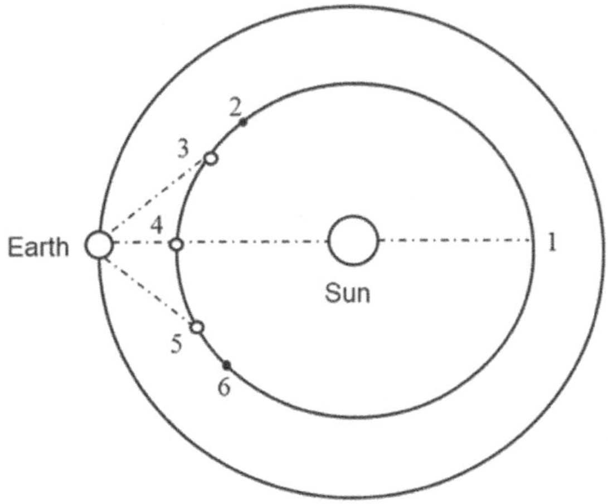

Figure 18. The retrograde motion of the inferior planets (The outermost circle is the orbit of the Earth, and the innermost circle is the orbit of the inferior planet.)

In the heliocentric diagram above, the inferior planet is in superior conjunction at point 1. At this point, the inferior planet reaches the same degree as the Sun's longitude. In this placement, which is called superior conjunction, we cannot see the inferior planet in the sky because the Sun is positioned in front of it. The inferior planet continues its direct motion around the Sun.

After that, the inferior planet arrives at point 2. In this placement, the inferior planet can be seen in the sky after sunset, and it continues along its orbit in direct motion. It begins to decelerate at point 2, and then it stops. At point 3, the planet switches to retrograde motion. In this placement, Mercury is 28° ahead of the Sun's zodiacal longitude while Venus is 48° ahead. This means that the inferior planet is at its greatest eastern elongation[54]. For instance, Venus is at its greatest eastern elongation 220 days after superior conjunction.[55]

54 A planet's "greatest elongation" is when it is at its farthest possible placement from the Sun.

55 If you look at the geometrical shape in Figure 23 formed by the orbital locations of Venus, you can see an isosceles triangle with an apex angle of 36° and base angles of 72° each. This triangle is called the "Golden Triangle." Venus adheres to this rule as it carries on with its motions.

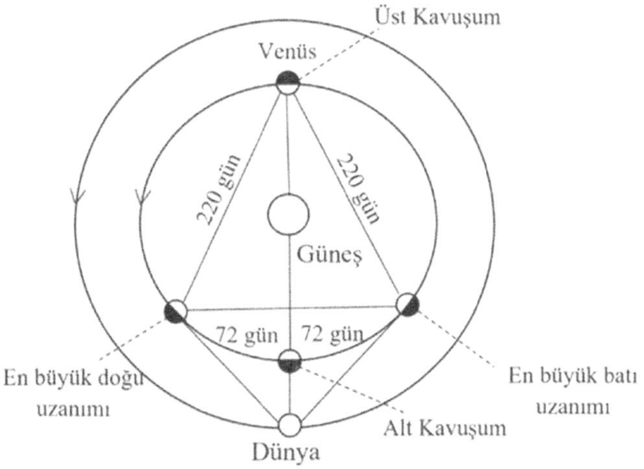

Figure 19. Superior conjunction, inferior conjunction, greatest western elongation, and greatest eastern elongation for Venus (an inferior planet)[56]

Mercury's average retrograde period is 22 days, and Venus's is 42 days.[57] At the midpoint of the retrograde period, the inferior planet is in conjunction with the Sun. After some time (this is about 10 days for Mercury and 21 days for Venus), at point 5, the planet appears to decelerate again, and then it continues its direct motion. The planet can be observed clearly from the Earth before station-retrograde and after station-direct, at points 2 and 6, or in other words, during its greatest elongations. At point 6, Mercury and Venus rise after the Sun and are called morning stars. At this placement, Mercury and Venus are located 28° and 48° behind the Sun's zodiacal longitude, respectively. At this period, the planet regains its direct motion.

56 Despite Venus's revolution period around the Sun, which is 225 days, it's synodic period is long—584 days—because of the closeness of its orbit to that of the Earth. Mercury completes one full revolution around the Sun in 88 days and passes through inferior conjunction every 116 days. The interval of time between the inferior conjunctions of these planets are called "synodic periods." (A synodic period contains not only conjunctions, but also the time elapsed between two consecutive oppositions.)

57 Mercury is in retrograde about 3 times (on occasion, 4 times) every 13 months, and Venus is in retrograde only once every 18 months.

Retrograde Motion Of The Superior Planets

Let's talk about the retrograde motions of the superior planets over a diagram. At point 1, the superior planet is in conjunction with the Sun, and the planet's apparent motion is at its fastest. At point 2, the planet's apparent motion begins to slow down. At point 3, the planet appears to have stopped. At point 4, it stops and begins its retrograde motion. At point 5, the Earth is situated between the Sun and the superior planet. So, the Sun and the planet are in opposition. However, if we were to observe from the Sun, we would see the Sun and the planet in conjunction. At point 6, the planet is in the fastest phase of its retrograde motion. At point 7, it decelerates. At point 8, the planet prepares for its direct motion.

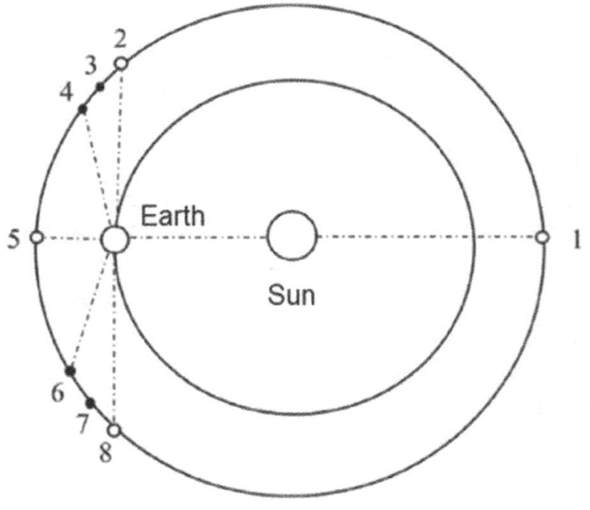

Figure 20. The retrograde motion of the superior planets

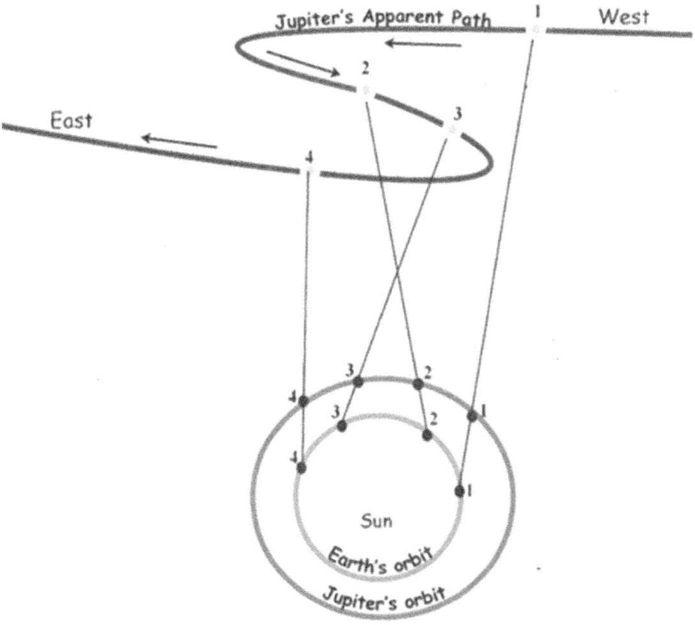

Figure 21. The retrograde motion of Jupiter

Now, let's talk about the retrograde motions of the superior planets individually.

Mars: Mars completes one full revolution around the Sun in 687 days. It is in retrograde for approximately 9% of this cycle. Mars revolves around the Sun at a speed of 24.1 kilometers per second, whereas this speed is 29.8 kilometers per second for Earth. Because of the speed difference, Mars appears to be in retrograde for 72 days every 25.6 months. The mean synodic period of Mars is 780 days.

Jupiter: Jupiter is in retrograde for approximately 30% of its yearly cycle. This corresponds to 110 days per year. The mean synodic period of Jupiter is 398.9 days.

Uranus: Uranus completes one full revolution around the Sun in 42 years. Uranus is in retrograde for approximately 41% of its yearly cycle, which corresponds to approximately 150 days of retrograde motion per year. The mean synodic period of Uranus is 369.7 days.

Neptune: Neptune's orbital period is approximately 165 years. It is in retrograde for approximately 43% of its yearly cycle, which roughly corresponds to 157 days. The mean synodic period of Neptune is 367.5 days.

Pluto: Pluto completes one full revolution around the Sun in 248 years. Pluto's roughly 17° inclination to the ecliptic results in variations in the time the planet spends in each sign. For instance, it spends 32 years in Taurus but only 11.5 in Scorpio, Taurus's opposing sign. Pluto is in retrograde for approximately 44% of its yearly cycle, which corresponds to about 156–161 days. The mean synodic period of Pluto is 366.7 days.

VIII. PHASES OF THE MOON

For an observer on Earth, the size of the illuminated side of the Moon varies periodically. This is caused by the Moon's revolution around the Earth. These observable changes in light are called "phases of the Moon" or "lunar phases." The Moon's phases are defined by a geometric sequence based on the Moon's and Sun's positions relative to each other. Based on the angle of incidence of the Sun's rays, the Moon has 8 basic phases that can be observed from Earth. Now, let's have a look at the four main phases of the Moon: new moon, first quarter, full moon, and last quarter.

The Moon cannot be seen during the **new moon phase** because the side of the Moon that faces us does not receive any light from the Sun. In the meantime, the Moon is situated between the Earth and the Sun. In the days that follow, the Moon begins to appear as a very fine crescent. This phase can be observed in the southwest at sunset. The crescent waxes gradually until it reaches the first quarter phase.

During the **first quarter phase,** the Moon appears as a semicircle on the right-hand side. This phase comes about 7.5 days after the new moon phase. The Moon moves away from the Sun by 90°, so it is directly overhead when the Sun sets. The first quarter phase can be observed in the south at sunset.

During the **full moon phase**[58], the side of the Moon that faces us is illuminated entirely, making this the brightest phase. There is a 14.5-day interval between a full moon and a new moon. During this phase, the

58 The full moon closest to the autumnal point is called the "harvest moon," because its brightness allows farmers to keep working through the night. This is usually the last full moon in September or the first full moon in October. If a second full moon occurs within the same calendar month, it is called a "blue moon." A blue moon occurs about once every 33 months.

Earth comes between the Moon and the Sun. As a result, the Moon can be seen in the eastern horizon just after sunset. After this phase, the Moon's light begins to wane.

During the **last quarter phase,** the side of the Moon that remains dark during the first quarter phase is illuminated; the bright semicircle is on the left-hand side. The angle between the Moon and the Sun is 270°.

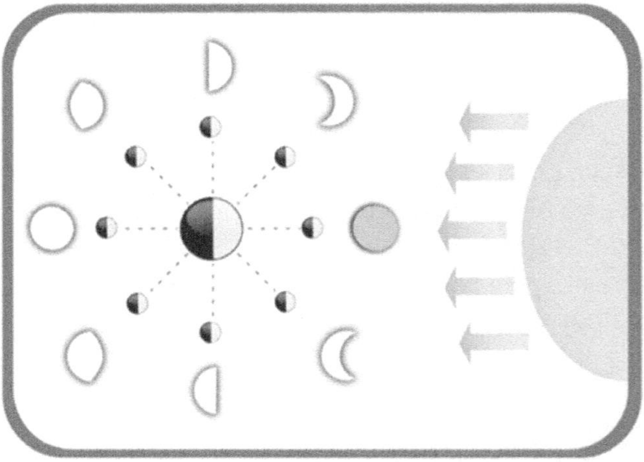

Image 20. Phases of the Moon

In order for us to see the same Moon phase again, the Moon must return to the same position with respect to the Sun and the Earth. The time required for a Moon phase to reoccur is called a "synodic month" or a "lunar month." One synodic month corresponds to 29 days, 12 hours, 44 minutes, and 2.78 seconds—in other words, 29.53 days. The time interval between two consecutive moonrises with respect to a star is called a "sidereal month," which corresponds to 27.32 days.

The Moon, like the Sun, participates in the daily movement by rising and setting, and it moves eastward by 13° every day in its orbit around the Earth. This is why the Moon shifts slightly with respect to the stars every day during its daily movement and rises 52 minutes later each day than the day before.

ECLIPSES

People of the early ages facilitated their lives by observing the movements of the Sun and the Moon. On some days, however, the sky deviated from its usual rhythm. The Sun abruptly cut off its light, and day became night. Our ancestors realized that eclipses did not occur randomly. The Babylonians figured out that lunar and solar eclipses occurred at regular intervals.

Back in the early ages, the time and location of eclipses held great importance. In fact, according to Chinese sources, which contain the earliest records of eclipses, astronomers who failed to predict an eclipse were killed. Solar eclipses were interpreted as bad omens in many civilizations. When an eclipse was near, the Babylonians and Assyrians would pick a temporary king. They assumed that this would help draw the wrath of the gods onto this temporary king. Having a temporary king who was killed after the eclipse allowed the prophecy to come true while securing the true king's sovereignty.

The Geometry Of Eclipses

In the geometry of eclipses, the alignment of the Earth and the Moon becomes significant as the Sun is fixed. Every day, the Earth moves eastward by 1° around the Sun. Consequently, the alignment of the Earth, Moon, and Sun changes constantly. This influences the manner in which eclipses occur. Eclipses occur when the Sun, Moon, and Earth are aligned on an almost straight virtual line. An observer on Earth sees two kinds of eclipses: solar eclipses and lunar eclipses.[59] In order for a solar or lunar eclipse to occur, the Sun and Moon must be on or near the lunar nodes.

59 As it happens, inferior planets become eclipsed as well. Transits of inferior planets across the solar disk qualify as eclipses. In principle, these eclipses occur in the same manner as solar and lunar eclipses.

What Are Lunar Nodes?

There is an angle of 5°8′ between the orbital planes of the Moon and the Earth. The orbits of the Moon and the Earth intersect at two points. Of these two points, the one at the north is called the "north node," and the one at the south is called the "south node."

Because of the irregularities in the Sun's apparent motion, lunar nodes shift westward by 3′ per day, 19° per year. Because of this shift, it takes 18.6 years for lunar nodes to return to the same point. Because of these motions of the nodes, the Moon does not remain above the horizon for the same amount of time on the same date every year. For instance, a node on the first point of Aries will have shifted westward by 19° a year later, moving below the equator, and so, the Moon will be above the horizon for less than 12 hours. So, one shouldn't expect to see the same phase on the same date every year. Since 12 synodic months add up to a total of 354 days, 11 more days are required to complete one year.

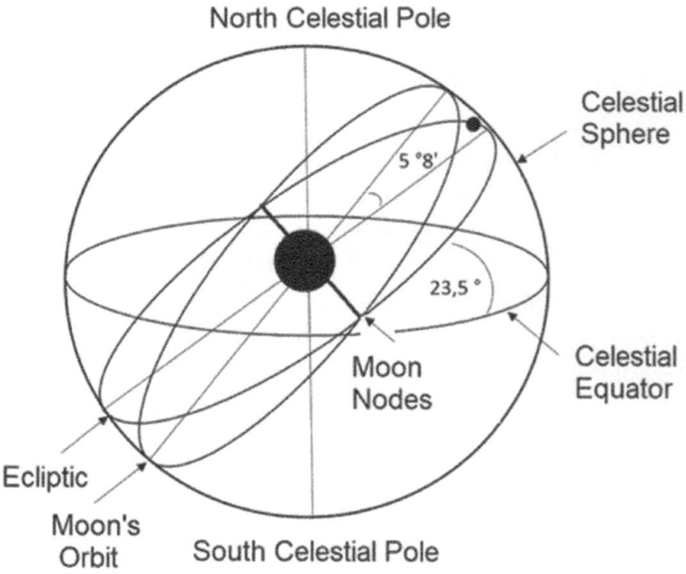

Figure 22. Lunar nodes

7 eclipses can occur in a year. The possibilities are as follows:

- 1 to 5 solar eclipses, and 2 lunar eclipses.
- 2 to 4 solar eclipses, and 3 lunar eclipses.

Once every two years, a partial solar eclipse occurs.

SOLAR ECLIPSES

Solar eclipses occur during the new moon phase, when the Moon is between the Earth and the Sun. However, not every new moon phase results in a solar eclipse. In order for a solar eclipse to occur, the Sun and Moon must be aligned with or near the lunar nodes.

Solar eclipses can be total, annular, or partial. Total and annular solar eclipses are determined by the Moon's distance from the Earth, whereas a partial solar eclipse requires the Sun, Moon, and Earth to be aligned.

During a solar eclipse, the Moon's shadow falls on the surface of the Earth. Since the Earth continues to spin in the meantime, the Moon's shadow traces a specific path on the Earth's surface. This path is known as the "eclipse path," and it can be as long as 16,000 kilometers and as wide as 160 kilometers.

The angular sizes of the Moon and the Sun are the same (about 0.5°).[60] However, during a total solar eclipse, the Moon is close to the Earth, so its apparent size[61] is slightly larger than that of the Sun. In this case, a total solar eclipse occurs. During a total solar eclipse, the white halo of light around the Sun expands and becomes discernible.

60 Surprisingly, the Moon and the Sun almost appear to be the same size in the sky. The Sun's diameter is 400 times that of the Moon. However, since the Moon is 400 times closer to us than the Sun, they appear to be the same size. Therefore, sometimes the Moon can fully obscure the Sun.

61 The Moon's distance from the Earth and the Earth's distance from the Sun are constantly changing. This causes their apparent sizes to fluctuate according to an observer on Earth. Consequently, sometimes the Moon appears larger than the Sun, and sometimes it appears smaller.

A total solar eclipse lasts 7.5 minutes at most. You must be somewhere along the eclipse path to watch a total solar eclipse. The Sun will appear completely dark to an observer in the umbra, but only a portion of the Sun will appear dark to an observer in the penumbra. Therefore, a total eclipse may appear as a partial eclipse to an observer elsewhere on Earth.

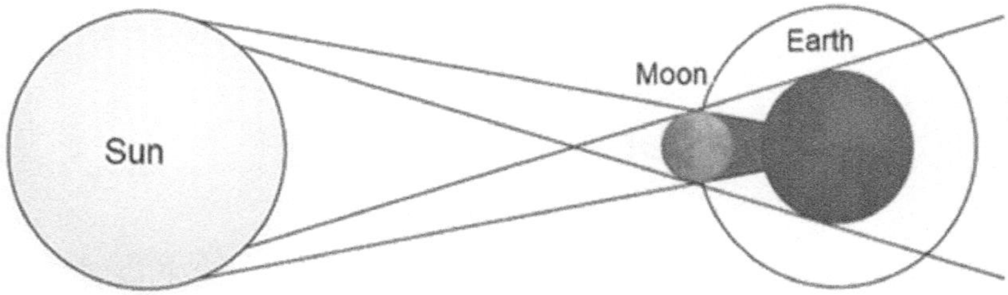

Figure 23. Solar eclipse

The innermost part of a shadow is known as its umbra, and the outer, partially shaded part is known as its penumbra. The relative areas of the umbra and penumbra are determined by the distance between the Moon and the Earth.

As it happens, we are a fortunate generation to have the opportunity to witness a total solar eclipse because, millions of years from now, the Moon will no longer be able to obscure the Sun. This is due to the fact that the Moon is drifting away from the Earth (the distance between the Moon and the Earth has increased 18-fold in 4.5 billion years[62]) and that the Earth's rotation is slower (the Earth's rotation has slowed by 2 seconds in 100,000 years). Generations who will not be able to witness total solar eclipses will only be able to witness annular solar eclipses.

Sometimes the Moon is at the farthest point of its orbit, causing it to appear smaller than its usual angular size. In such cases, the Moon cannot obscure the Sun completely. In instances where the Sun's apparent size is larger than that of the Moon, an annular solar eclipse occurs. Observers see annular solar eclipses as a ring of light. The Sun's outline can be seen around the Moon. In this case, the Moon's umbra does not reach the Earth.

62 When the Moon first came into existence, its distance from the Earth was 22,530 kilometers, whereas now, this distance is 384,500 kilometers on average.

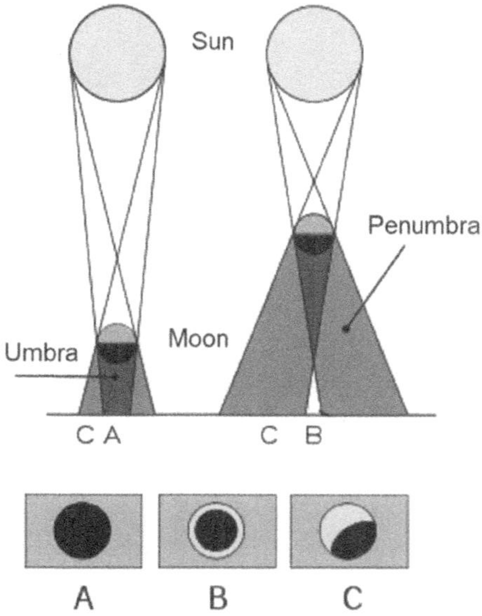

Figure 24. Types of solar eclipses

During a partial solar eclipse, the Moon obscures a portion of the solar disk. Solar eclipses observed from points within the Moon's penumbra will appear as partial solar eclipses. In fact, partial solar eclipses can be seen from certain parts of the Earth during both total and annular solar eclipses.

Partial eclipse phases may also be witnessed before and after total and annular eclipses occur. Sometimes, a solar eclipse may begin as an annular eclipse, progress to a total solar eclipse, and then return to an annular solar eclipse. The partial eclipse phase of all eclipses lasts quite a while.

If the new moon is between 0° and 9°55′ from the lunar nodes, a total solar eclipse occurs. If the new moon is between 9°55 and 11°15′ from the lunar nodes, a partial or total solar eclipse occurs. If the new moon is between 11°15′ and 15°21′ from the lunar nodes, a partial solar eclipse occurs. If the new moon is between 15°21′ and 18°31′ from the lunar nodes, again, a partial solar eclipse occurs.

Looking at the Sun with the naked eye or through binoculars during an eclipse may damage the retina, thus harming the eyes. Therefore, it is crucial to use the appropriate filters to observe a solar eclipse.

LUNAR ECLIPSES

Lunar eclipses occur during the full moon phase. The Earth comes between the Moon and the Sun, preventing the Sun's rays from reaching the Moon. Just as there isn't a solar eclipse every new moon, there isn't a lunar eclipse every full moon. This is due to the incline of the Moon's orbit with respect to the Earth's orbital plane. A lunar eclipse can only occur when the Moon is in its full phase or is near the lunar nodes.

Here is the basic difference between a lunar eclipse and a solar eclipse: during a solar eclipse, the Sun's light is obscured by a tangible celestial body (the Moon), whereas lunar eclipses are caused by the Earth's shadow shrouding the Moon.

The area where the Earth inhibits not all, but rather a certain portion of the Sun's light from reaching the Moon is called the "penumbra" or "partial shadow." The "umbra," or the fully shaded area, is where the Earth completely inhibits the Sun's light from reaching the Moon.

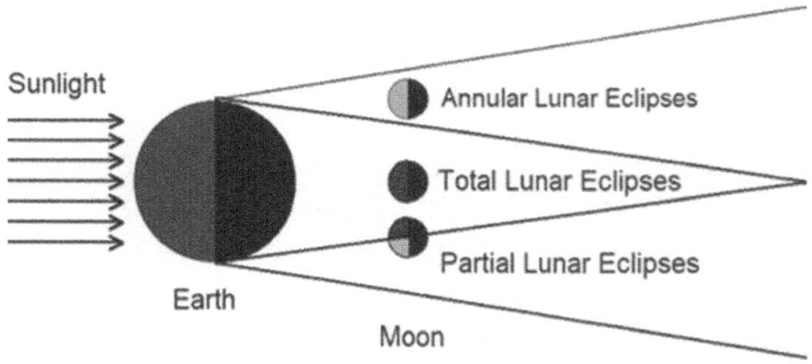

Figure 25. Types of lunar eclipses

The shadow that the Moon casts over the Earth during a lunar eclipse is several times larger than the Moon's diameter. Since the Moon is at its closest to the Earth, it appears about 7% larger than other full moons during a lunar eclipse.

A total lunar eclipse occurs when the Moon completely passes into the Earth's umbra. The Earth's atmosphere filters and refracts the sunlight during a total lunar eclipse, creating an interesting and beautiful sight.

If it weren't for the Earth's atmosphere, the Moon would be pitch black during a total lunar eclipse. However, during a total lunar eclipse, the Moon can appear to take on a variety of colors, ranging from dark brown and red to bright orange and yellow. The exact color of the Moon's surface depends on the amount of dust in the Earth's atmosphere.

A total lunar eclipse can last over an hour (about an hour and 40 minutes). This is the amount of time the Moon spends in the umbra. Since the Earth's shadow is large enough to shroud the Moon completely, lunar eclipses can be observed anywhere it is nighttime at the time of the eclipse. Lunar eclipses appear the same from any location from which they can be observed, which means that both the eastern and western sides of a country get the same view during a lunar eclipse.

The Moon is too far away to cast an umbra during a penumbral lunar eclipse, so an actual total eclipse does not occur. A partial lunar eclipse occurs when only a portion of the Moon enters the Earth's umbra. In this case, only a portion of the lunar disk becomes invisible. A partial lunar eclipse is quite difficult to see with the naked eye. Since the part within the Moon's penumbra can only receive light from one portion of the Sun at that moment, it appears dimmer than usual.

If the full moon is between 0° and 3°45′ from the lunar nodes, a total lunar eclipse occurs. If the full moon is between 3°45′ and 6° from the lunar nodes, a partial or total lunar eclipse occurs. If the full moon is between 6° and 9°30′ from the lunar nodes, a partial lunar eclipse certainly occurs. If the full moon is between 9°30′ and 12°15′ from the lunar nodes, a penumbral lunar eclipse may occur.

SAROS CYCLES

One can predict ahead of time the date of an eclipse, as well as where it can be observed. The Babylonians kept track of astronomical events, and they discovered that eclipses occurred about once every 18 years. In the present day, we call this the "saros[63] cycle." Saros cycles were

63 The word saros is derived from the Babylonian word sāru, which refers to 3600-year periods.

also known by Hipparchus, Pliny, and Ptolemy, who lived after the Babylonians' era.

A saros cycle is an eclipse cycle that is 18 years, 11 months, and 8 hours (about 6585 1/3 days) long. It is used for predicting solar and lunar eclipses. After a saros cycle that follows an eclipse, the Sun, Moon, and Earth align in the same manner with respect to each other, and a similar eclipse occurs.

Each eclipse within a series appears slightly different from the others. Each eclipse occurs about 18 years later in the same sign, with an approximately 10° shift. In that case, if we add this amount of time to the date of an eclipse, we will see another eclipse on the resulting date. Because the orbital motions of the Sun and Moon differ, this eclipse will not occur at the exact same time as the previous one. Again, calculations show that 54 years and 34 days after the initial eclipse, another eclipse occurs.

A saros series begins with a partial eclipse, and in each successive saros cycle, the Moon's path shifts north-ward (when near the descending node) or southward (when near the ascending node) compared to the previous one. Each eclipse series in a saros cycle begins with a partial eclipse at one of the poles, followed by a total eclipse. Then the series moves away from the lunar nodes, resulting in a partial eclipse. At some point, eclipses are no longer possible, and the series ends. The series ends at the opposite pole to the one at which it began. This process takes 1300 years.

A saros series is numbered based on whether the eclipse is solar or lunar, and whether it occurs near the ascending or descending node. Solar eclipses that occur near the ascending node are given an odd number, while those that occur near the descending node are given an even number. This numbering system is reversed for lunar eclipses.

IX.
THE STARS

You can see thousands of stars with the naked eye and millions of stars with a telescope. The stars are trillions of kilometers away from us. At a distance of 149,637,000 kilometers, the Sun is the star that is closest to the Earth. The second closest star is Proxima Centauri[64]. Proxima Centauri is approximately 4 light-years (1 light-year corresponds to 9,467,280,000,000 kilometers) away from the Sun. Here is how we can express these distances using everyday measurements: let's say that the distance between the Sun and the Earth is one centimeter. In that case, Proxima Centauri's distance from the Sun would be over 40 kilometers. Thus, Proxima Centauri is too faint to be seen with the naked eye. On the other hand, stars such as Sirius, Pollux, Arcturus, and Aldebaran shine brightly in the sky because of their size despite being further away from the Sun.

The Classification Of Stars

Stars are classified according to their observable qualities. Most of them are Sunlike stars. Some, however, are quite large. These are called "giants" or "supergiants." The smaller ones are called "white dwarfs."

Most stars that appear as solitary stars in the sky are actually binary, trinary, or even quaternary star systems. If the system consists of two stars, it is called a "binary star." If it consists of three stars or more, it is called a "multiple star." The Sun is among the very few stars that are solitary.

64 Proxima Centauri is a red dwarf star. Its size and mass are about 1/7 of those of the Sun. It gets its name from its reddish color.

The greater the mass of a star, the shorter its life span will be, and the smaller the mass, the longer the life span. For instance, the Sun is a small star. According to calculations, it is about 4.5 billion years old and is expected to live for another 5 billion years. A star 100 times larger than the Sun would have a lifespan of approximately 100 million years.

Stars come in a range of colors. Stars with large masses are more bluish and brighter. Stars with smaller masses are more reddish and fainter. Since the color of a star is directly proportional to its temperature, we can detect a star's surface temperature by detecting its color.

The Brightness Of Stars

When observing the stars with the naked eye, some stars appear brighter than others. When we talk about the brightness of a star, we actually refer to the effect that that star has on our eyes. A star's brightness is determined by its distance from the Earth and its luminosity.[65]

In 150 BC, Hipparchus prepared a catalogue of the locations and brightnesses of 1080 stars. The brightness of the stars is measured by a unit called "magnitude." Hipparchus called the brightest stars "firstmagnitude stars" and the faintest stars visible with the naked eye, "sixth-magnitude stars." A first-magnitude star sends us 100 times more light than a sixth-magnitude star does. Magnitude is denoted by a superscript m. So, a first-magnitude star would be denoted by "1^{m}." Bright stars may have negative magnitude values.

If you get away from the city lights, you can see the fifth-magnitude and sixth-magnitude stars. Because of telescopes, we now know that there are stars fainter than the sixth magnitude.

Some stars' brightnesses don't remain constant, but change over time. These stars are called "variable stars." Although there have been some alterations in the present day, the brightnesses of the stars are still based on Hipparchus's classification.

65 As the distance between an observer and a star increases, the star's brightness decreases proportionally to the square of the distance.

Naming The Stars

The stars that comprise the constellations are named based on their brightness. The brightest star of a constellation is given the first letter of the Greek alphabet, alpha (α). The others are given letters such as beta, gamma, delta, or epsilon based on their order of brightness. The Greek letter is followed by the name of the constellation. For instance, "α Canis Majoris" is the brightest star in the constellation Canis Major.

Image 21. The alphabet used for naming the stars

Star Or Planet?

We can tell whether a light that we see in the sky is a star or a planet. First, let's go over the differences between stars and planets: stars produce their own energy, whereas planets reflect the light of a nearby star. Stars are scattered all over the sky, and their locations don't change. Planets are found in the belt where the Sun's path across the sky is situated. Their locations in the sky change constantly.

Although most books suggest that stars shine and planets do not, this is entirely incorrect. Planets appear more matte and disk-shaped, but they also appear to twinkle when they are close to the horizon. Their brightness is relatively lower, but not non-existent. Twinkling is associated with the rattling of light caused by the Earth's turbulent atmosphere. Furthermore, as planets' distances from the Earth change, their apparent brightnesses and angular sizes change as well. The closer a planet is to the Earth, the brighter and larger it will appear. Large planets appear disk-shaped when viewed through a small telescope or a pair of binoculars, whereas stars appear as dots even when viewed through the largest telescope.

Not every star has the same brightness. You can find even the faintest stars using a sky atlas.[66] However, while evaluating the brightness of the stars, the influence of atmospheric absorption must be considered. Comparing the apparent brightnesses of two stars of approximately the same brightness, one near the horizon and the other near the zenith, should suffice. The effects of twinkling are less pronounced near the zenith than near the horizon.

66 A sky atlas is a map used for determining the locations of celestial bodies in the sky. Sky atlases are created in a manner similar to maps of the surface of the Earth.

X.
CELESTIAL OBSERVATION

"...constancy lives in realms above
And life is thorny, and youth is vain"

-Samuel Taylor Coleridge, *Christabel,* 1800

The easiest way to observe the sky from a broad perspective is to observe it with the naked eye. With the naked eye and no observation tools, you can see the Moon, the constellations, and the motions of the planets; and with a sharp sight, you can see the phases of Venus, lunar and solar eclipses, meteor showers, and even the Andromeda galaxy, which is millions of light years away. You can do even more with a pair of binoculars or a telescope; consider how blurry everything appears when you dive into the sea without swimming goggles, and how clearly you can see everything when you put on a pair of swimming goggles and dive back in. Similarly, using telescopes and binoculars allows you to observe celestial bodies that you can't see with the naked eye.

On a dark night with no city lights, you can see about 4000 to 6000 stars when you lift your head up and look at the sky. If you are looking at the sky from a place where light pollution is prevalent, you can see no more than 500 of these stars. At this point, the concerns of when and where to observe the sky become significant.

The best observation you can make during the day is of the Sun. Various observations can also be made after sunset, until it gets dark, i.e., at twilight. Nighttime celestial observations are usually conducted on nights when the sky is clear and there is no Moon (although one can also choose days close to the first quarter phase). Another point that is worth noting is the observational quality of the atmosphere. If you stay in the dark for about 5 to 10 minutes before you begin your observation, your eyes can adjust to the dark. This way, you can even see the fainter stars in the sky.

In order to observe the sky, you must look in the right direction at the right time. Determining the positions of celestial bodies is important in observational astronomy. Needless to say, this requires a good knowledge of the sky. You can begin your celestial observation by acquiring a sky atlas and a red flashlight, which is the best at adapting to darkness. If possible, get a simple pair of binoculars. Try to start with the more prominent constellations. If you're patient, you'll discover that you can see more and more celestial bodies as time goes on. Now, let's look at some practical methods that can be of use to us during celestial observations.

Measuring Degrees By Hand

Measuring degrees by hand is one of the practical methods used for celestial observations. Astronomers define the distances in the sky, the area covered by the constellations, and the sizes of celestial bodies angularly. What we know is that the distance from the horizon to the zenith is 90°. Here is how we can measure this: extend your arm straight ahead and make a fist with your hand. Your fist covers an arc of about 10° in the sky. If you put one fist on top of the other nine times, you will reach the zenith. If you don't get that exact result, then your fist isn't exactly 10°, but it's close. It should be noted that these measurements are not the same for everyone; some minor variations among individuals are to be expected. Through regular observations, you can create your own measuring rules as well.

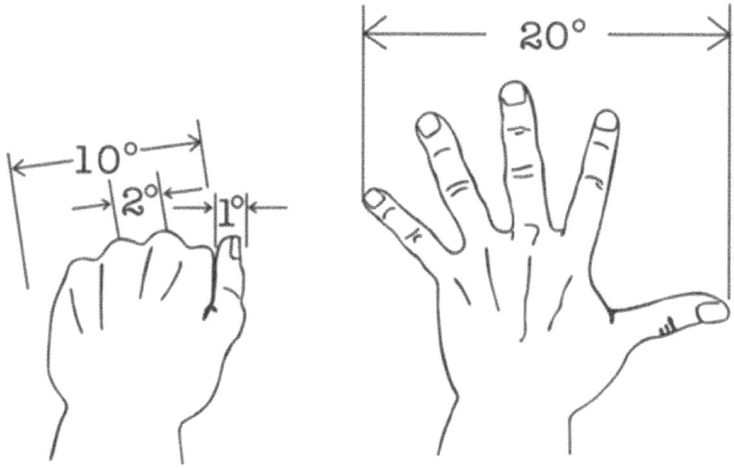

Image 22. Measuring degrees by hand

Even these approximate measurements can be used to detect the positions of certain celestial bodies quickly and effectively.

You can also measure degrees with your fingers. Your hand span is around 20°. The tip of your pointer finger is about 1° wide. Using these measuring rules, you can form statements such as "the star 20° (one hand-span) west of the Moon" or "the star 13° (a little more than a fist) south of Mars."

How Do We Determine The Directions?

Determining the directions is another key aspect of celestial observation. The Pole Star is oriented northward. If you turn your back to this star and spread your arms to both sides, your right hand will be pointing east, and your left hand will be pointing west.

We can use the constellation Ursa Major as a reference in order to find the pole star. Ursa Major is a constellation that can be easily found when observing from a spot with clear skies and no light pollution. Also known as the Big Dipper, Ursa Major consists of seven bright stars, as shown in the image. The two stars at the front edge of Ursa Major are

Merak and Dubhe, also known as pointer stars. An imaginary line joining these two stars points to the pole star. First, measure the angular distance between these two stars with your hand. Then, move roughly five times the distance you obtained in the previous step toward the open end of the dipper, in the same direction as the pointer stars. You will reach the Pole Star, or Polaris[67].

Other Methods

Using a compass is the easiest way to find one's direction. The dark-colored end of the compass needle will be pointing north. On compasses, north is denoted by the letter N, and south by the letter S. We can also orient ourselves using the watch method: on a sunny day, turn the hour hand of your wristwatch toward the Sun. The bisector of the angle between 12 o'clock and the hour hand will indicate the south-north axis. The sunny side points south. We can also use nature to orient ourselves. The mossy sides of trees face north.

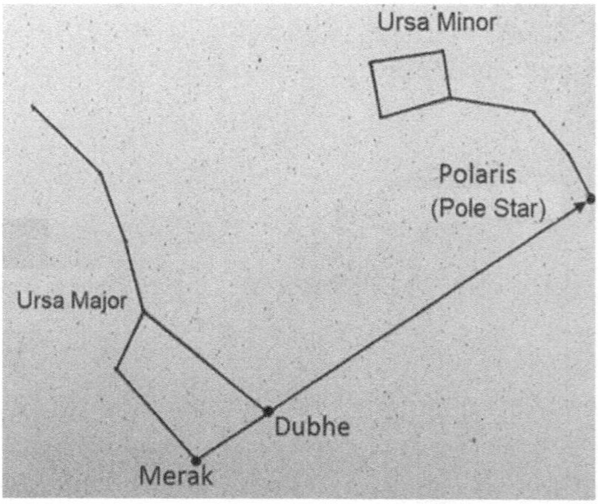

Image 23. Finding Polaris by using the Ursa Major as reference

67 Precession causes the pole star to change. Thuban, the brightest star in the constellation Draco (Dragon), was the pole star in 3000 BC. Gamma Cephei (Alrai) will be closer to the celestial pole point than Polaris in 3000 AD. Around 5200 AD, Iota Cephei will be the pole star. On 13,000 AD, the Earth's axis will again point toward Vega, and Vega will once again be the pole star. Polaris will be the pole star again on 27,800 AD, but it will be further away than it is now.

Observing The Sun

When we hear the phrase "observing the sky," we immediately think of nighttime observations. However, if we are talking about observing the Sun, we must do so during the day. The Sun's position on the celestial sphere changes every day due to its yearly apparent motion. As a result, the points at which the Sun rises and sets change every day.

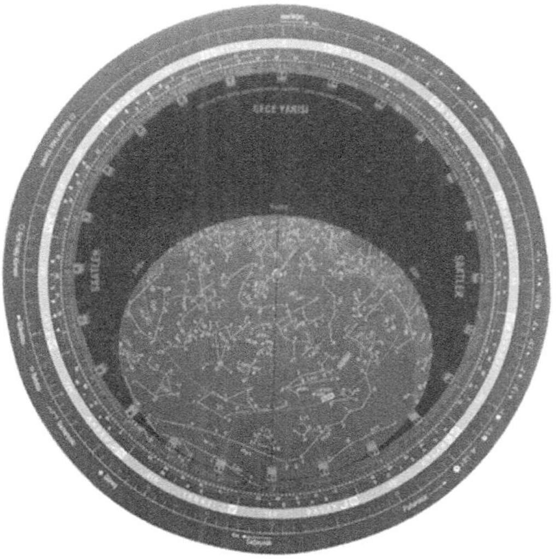

Image 24. Sky atlas

The angular size of the Sun is 0.5°. Let's demonstrate this with a simple experiment: let's say that the Sun is getting in your eyes and you want to block it with your hands. What would you do? Most likely, you would hold up your open hand (and thus, your fingers) to the Sun. However, you could block it out with just half of your thumb. Considering that one finger is 1°, half a finger will correspond to 0.5° of angular distance. A diameter of this size will be enough to see the details on the surface of the Sun through a telescope.

Since sunbeams hold an immensely powerful energy, you should never look at the Sun with the naked eye. Even looking at it through a telescope could damage your eyes. Therefore, it is customary to either use a special filtered telescope or project the image obtained through the telescope onto a screen. Although the latter doesn't provide us with the best view of the details on the Sun's surface, it is, none-theless, one of the methods of solar observation. Both these methods allow us to observe the surface of the Sun—i.e., the photosphere. You can also observe the sunspots. If you observe them regularly, you will notice that their shape and size fluctuate. You may recall that Galileo discovered that the Sun rotated around its own axis by observing the movements of sunspots.

Observing The Moon

The Moon is one of the celestial bodies that can be observed easily with the naked eye. This is because of its closeness to our Earth. The stars appear as dots when observed through binoculars or a telescope, whereas the surface features of the Moon can be seen in great detail with the naked eye.

If you look at the sky at the same time every day, you will notice that the Moon isn't always in the same spot. This is because the Earth's rotation around its axis and the Moon's revolution around the Earth are in the same direction. After the Earth makes its diurnal motion around its own axis, it needs to rotate for an additional 50 minutes to catch up to the Moon. This is why the Moon rises 50 minutes later and moves eastward by 13° every day in comparison to the previous day. In fact, the Moon rises so late some nights that it is only possible to observe it early in the morning.

One of the most enjoyable aspects of lunar observation is watching the Moon's surface features. The darker zones are the lunar maria, and the brighter zones are the craters. With the naked eye, you can only see the maria, some of the more prominent craters, and the highlands, whereas a small telescope allows for more detailed observations. Watching the phases of the Moon also provides us with a visual feast. However, it is so bright in the full moon phase that it causes discomfort to the eyes when viewed through a telescope or binoculars. At this phase, the rays strike the surface perpendicularly, causing shadows to disappear. This makes it difficult to distinguish many surface features.

The Moon doesn't have its own light. It reflects an average of 7% of the light emitted by the Sun. Nonetheless, it illuminates the sky brightly enough to inconvenience observers. The brightness of the Moon significantly reduces the number of celestial bodies that can be observed in the sky. Therefore, many novice observers avoid observing the sky on nights when the Moon's light is bright.

Observing Mercury

Mercury is so close to the Sun that seeing it requires a clear horizon and careful observation. With a magnitude that can reach −1.9, its brightness is close to that of Sirius—in fact, sometimes it becomes even brighter than Jupiter.

Mercury's greatest angular distance from the Sun is 28°. However, even at a distance of 28°, it is difficult to observe Mercury. It can rarely be seen with the naked eye. If you want to look at Mercury using binoculars or a telescope, you must wait for the Sun to set completely; if the Sun accidentally enters your field of view, it can damage your eyes.

Mercury can be seen six times a year as an evening star and once a year as a morning star. It can be observed for two hours after sunset in March and April, and before sunrise in September and October.

Transits Of Mercury

Mercury's orbit is inclined by 7° to the Earth's orbit. Therefore, Mercury doesn't always come inside the solar disk. It can be seen in the solar disk only when it is in conjunction with the Sun at the same time that it crosses one of the nodes where the planet's orbit intersects the ecliptic. Roughly 13 transits of Mercury can occur in each century. A full transit occurs in the months of May and November. The previous transits of Mercury occurred on November 6, 1993; November 15, 1999; May 7, 2003; and November 8, 2006. The next transits of Mercury will occur in November 2019 and November 2032. Since Mercury is too small to be seen with the naked eye during transit, it can only be observed through a telescope. It's also a good idea to compare Mercury to any sunspot in one's field of vision; this way, it becomes evident that Mercury is pitch black, whereas the spot isn't.

Observing Venus

Venus is the brightest object in the sky after the Sun and the Moon. Its brightness ranges between magnitudes –3.3 and –4.3. This is about ten times the brightness of the brightest star in the sky.

Venus can be observed even when the Sun is above the horizon. It is best observed during the 5.5-hour period before sunrise or the 5.5-hour period after sunset. Venus, like Mercury, moves in close proximity to the Sun. Its maximum elongation is 48°. Apart from its times of inferior and superior conjunction with the Sun, it is easily visible. To locate Venus in the sky, you must know its approximate position in relation to the Sun. If you intend to observe Venus with binoculars or a telescope, avoid looking directly at the Sun.

If you observe Venus with a pair of binoculars, you can see that it has phases like those of the Moon or Mercury. When it is closest to Earth—that is, when it is in inferior conjunction—it presents to us its dark side. In that case, we don't see Venus except during its transits, which are rare. When Venus is in superior conjunction, which is when the planet is behind the Sun and at its furthest point from the Earth, its visible side is entirely illuminated, and the planet is in full phase, appearing as an illuminated disk. During this phase, Venus is in apogee, and its apparent size is the smallest. Therefore, it is not bright enough. When Venus is at maximum elongation, it appears as a semicircle. In the crescent phase, it is at its closest to the Earth and its diameter is the largest. However, the Sun's beams prevent Venus from being seen. Venus's apparent brightness is at its highest (30%) between its crescent and half-illuminated phases.

Even if you observe Venus through a powerful telescope, you won't see much more than a bright circle. Its surface appears silver-colored and spherical, and there are faint, ambiguous shapes on it. These shapes move around swiftly. These aren't surface features of Venus; they are clouds in the upper layers of its atmosphere. These clouds are responsible for Venus's bright appearance: the light reflecting off the clouds gives the planet its striking bright white look. As Venus transits the Sun, it leaves behind a black trace. This trace disappears sometime after the beginning of Venus's transit. This effect is, again, caused by Venus's atmosphere.

Transits Of Venus

Although Venus passes through inferior conjunction about every 20 months, it rarely transits the solar disk because of the 3.39° incline of its orbit to the plane of the ecliptic. In order for Venus to transit the solar disk, the points where its orbit intersects the plane of the ecliptic must be aligned with the Sun and the Earth. Transits of Venus occur approximately twice each century in pairs separated by 8 years (1761-1769, 1874-1882, 2004-2012, 2117-2125). When these transits are analyzed in further depth, it can be seen that they occur in a 243-year cycle. All transits occur in the months of June and December, which coincide with the nodes' projections on the Earth's orbit. A transit of Venus appears as a small, dark dot traveling across the solar disk and lasts no more than 7 hours.

Observing Mars

Mars appears as a reddish-orange dot of light to the naked eye. Its brightness increases as its position relative to the Earth changes; it reaches magnitude –2.9. Its brightness is more than 50 times greater when it is closest to the Earth than when it is farthest away, just like a star.

Like Venus and Mercury, Mars has phases as well. It cannot be seen from Earth during its half and crescent phases. It can be observed in those phases that are similar to the lunar phases before and after a full moon. When Mars is in opposition—that is, when the Earth is between the Sun and the planet—it rises at sunset and remains in the sky throughout the night.

Mars is not so easy to observe. First of all, it is small; its diameter falls between that of the Moon and the Earth. If you observe Mars with binoculars, you can see its color and brightness more clearly. When observed with a large telescope, you may not notice its surface features at first, but as you become more experienced at observing, you can make out the colors and shapes on its surface.

Observing Jupiter

After the Moon and Venus, Jupiter is the brightest visible celestial body in the sky. Its brightness exceeds that of Sirius for the majority of the year, reaching a magnitude of –2.7. It is easily visible with the naked eye during the one or two-month period when it is in opposition with the Sun and at its closest to the Earth.

For a novice observer, Jupiter is a planet that is worth observing. If you observe it through a telescope, you can see dark and light-colored bands on it. Despite its size, the Great Red Spot is difficult to observe. The spots on the outer surface of Jupiter move constantly. It takes at least six hours for a spot to move from one side of Jupiter to the other.

You can observe Jupiter, as well as its four satellites, with a simple pair of binoculars. However, since the satellites move, the sight you see will change every day; sometimes they may move behind Jupiter, and sometimes they may move in front of it. These satellites have roughly the same brightness.

Observing Saturn

Saturn provides us with one of the most beautiful sights we can see in the sky. It appears faint since it is farther away from the Sun than it is from Jupiter and smaller than it as well. With its yellowish color, the highest brightness it reaches is magnitude –0.3. This means that Jupiter is 10 times brighter than Saturn. Titan, one of Saturn's satellites, has a magnitude of 8 and can be observed with a small telescope.

Saturn's rings move throughout the years because of the planet's axial tilt. Saturn's rotational axis makes an incline of 26.73° to its orbital plane. As a result, as the planet moves in its orbit, the plane of its rings appears to shift when viewed from Earth. Because the appearance of its rings varies, so does their brightness. Saturn's brightness is increased by the light reflecting off of its rings; therefore, Saturn appears brighter during the years when its rings are facing us.

The Earth passes through Saturn's ring plane twice in one orbital period of Saturn, which is around 29.5 years. In this case, its rings cannot be seen. For example, Saturn's rings were invisible in 1995 and 2009, appearing only as a bright line passing through the planet's center. The same thing will occur in the year 2037.

Observing Uranus

Uranus has a magnitude of 6, which is why it can only be observed under extraordinarily clear conditions, and even then, it appears as a small, faint star. This is due to its remoteness as well as its small size. With a small telescope, it can be seen as a blue-green circle, albeit without detail. Its satellites can only be observed with very large telescopes.

Observing Neptune

At a magnitude of 8, Neptune is quite faint. This is caused both by its remoteness and its small size. This is due to its remoteness as well as its small size. It is not visible to the naked eye. When observed with binoculars or a telescope, it appears as a matte blue spot. Triton, its satellite, can be observed under excellent conditions with a 20-centimeter telescope.

Observing Pluto

Pluto's average magnitude is 14. Even with a telescope, it can only be seen as a small dot of light.

XI.
CONSTELLATIONS

Stars far apart in the sky appear to us to be very close to each other and at the same distance from the Earth. This is due to our eyes' inability to perceive depth at great distances.

Stars appear as if they are creating distinctive shapes. People of the antique era named these star groups based on their shapes, appearances, and locations, thus creating the constellations. They divided the sky into sections by stars with well-defined borders, much in the same way we divide a map of Turkey by cities.

The constellations were named after mythological figures (such as Hercules, Perseus, and Zeus), animals (Tucana, Canis Major, and Canis Minor), inanimate things, or other themes. For instance, they named a constellation "Orion" (the hunter) because they thought that it resembled a hunter. This made the constellations easier to remember. The constellations facilitated determining the locations of the celestial bodies in the sky.

Sky atlases, which are similar to maps of the Earth's surface, were developed over time to aid in the detection of celestial bodies' locations in the sky. In the 2nd century AD, Ptolemy created a sky atlas based on the catalogue created by Hipparchus. In this sky atlas, he identified and named the 48 constellations visible at his latitude.

In the present day, the boundaries of the constellations are defined by the International Astronomical Union (IAU). The IAU has identified 88 constellations in the sky in a manner that reflects the heritage of the past. 44 of these constellations are in the northern hemisphere, and 44 of them are in the southern hemisphere. Many of these constellations' names that are in use are still the same as they were on Ptolemy's lists. Twelve of these are located on the ecliptic and are the origins of the zodiac signs' names.

The Constellation Aries

Figure 26. The constellation Aries

The constellation Aries contains many binary stars and various galaxies that can be seen with telescopes larger than 30 centimeters. There are three stars in the constellation that can be seen with the naked eye: Hamal, Sheratan, and Mesarthim. Hamal is the brightest star in the constellation Aries. It has a magnitude of 2.2 and is 66 light-years away from the Earth. The constellation Aries can be observed best during autumn, in the month of November, in the northern hemisphere.

The Constellation Taurus

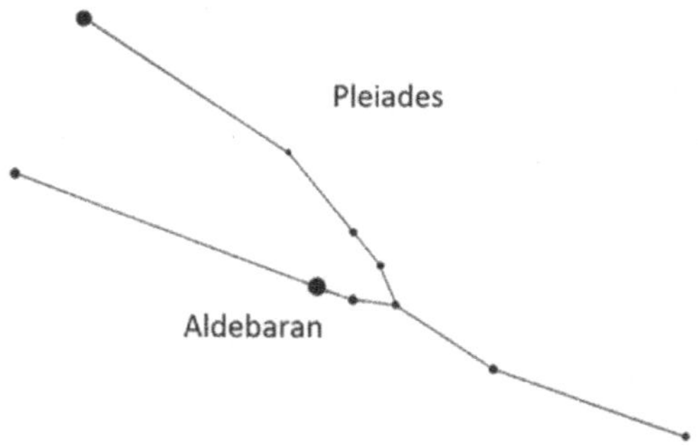

Figure 27. The constellation Taurus

The brightest star in the constellation Taurus is Aldebaran. Aldebaran is a red giant that is 65 light-years away from us. It is the 14th brightest star in the sky. Aldebaran and the stars near it form a small "V" shape.

The star clusters in this constellation that can be seen with the naked eye are Hyades and Pleiades. With the naked eye, the Pleiades star cluster appears as six stars arranged in the shape of a very small dipper. The Pleiades is a remarkable group; also known as "the Seven Sisters," as "Ülker" in Turkish, and as "al-Thurayya" in Arabic; it is 400 lightyears away. There are approximately 200 stars in the Hyades star cluster. It is 130 light-years away from the Earth. Although Hyades can be seen with the naked eye, it doesn't fit into the telescope's field of view all at once since it is spread out over a large area.

Furthermore, the constellation Taurus has nebulae that are invisible to the naked eye. The most popular of these is the Crab Nebula. This nebula, which is a supernova remnant, has a magnitude of 8.4. In order to observe it, one would need to look toward the horns of the Bull with a telescope.

Taurus is a large constellation that particularly stands out in the winter sky in the northern hemisphere. It can be best observed in the months of December and January.

The Constellation Gemini

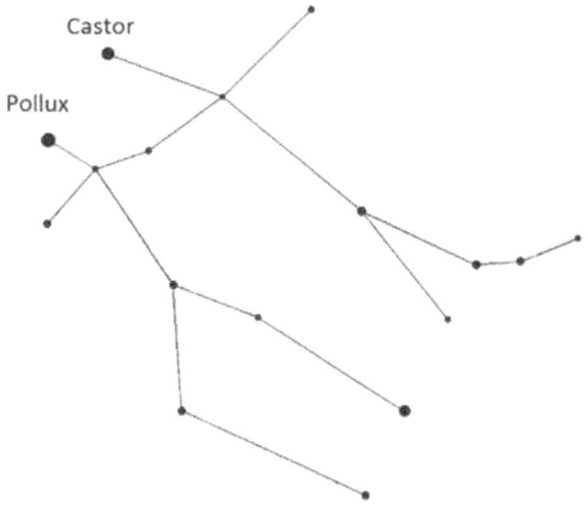

Figure 28. The constellation Gemini

Gemini is a constellation near the Milky Way, which is why it is full of bright stars. The brightest stars in the constellation are Pollux, with a magnitude of 2.7, and Castor, with a magnitude of 3.7. They comprise the head of the constellation Gemini. Castor is located 4.5° southeast of Pollux. Among all the binary stars in the constellation, Castor is the most prominent.

The constellation Gemini is home to many galaxy clusters. Within the boundaries of the constellation Gemini, there is the M35 open star cluster, among the group of stars beneath Castor. This cluster, which is 2800 light-years away, can be seen with the naked eye under excellent observation conditions.

The constellation Gemini can be best observed in February. It can also be seen in the west after sunset in April and May.

Finding Castor and Pollux Through Ursa Major

The constellation Ursa Major (also known as "The Big Dipper") can be used to find the constellation Gemini. Imagine a line passing through the two stars—Phecda and Merak—that form the bottom of the dipper's bowl and extend this line about 50° in the opposite direction to the dipper's handle; you will find the two stars about 8°–10° apart from one another. These stars are Castor and Pollux, which are part of the constellation Gemini.

The Constellation Cancer

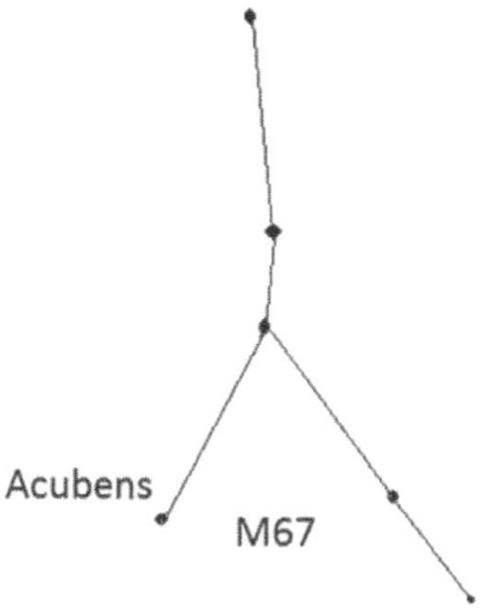

Figure 29. The constellation Cancer

There are very few stars in the constellation Cancer. Its most notable feature is a nebula.

The constellation Cancer has two open clusters. The first of these is Praesepe, which can be seen as a bright smudge of light with the naked eye. Galileo was able to observe 36 stars in this cluster with his telescope. In the present day, we can see more of its stars with modern telescopes. The other star cluster, M67, has 200 stars with magnitudes ranging from 10 to 15. It is not as easily observed as Praesepe.

The constellation Cancer can be observed at the beginning of spring in the northern hemisphere and in autumn in the southern hemisphere.

The Constellation Leo

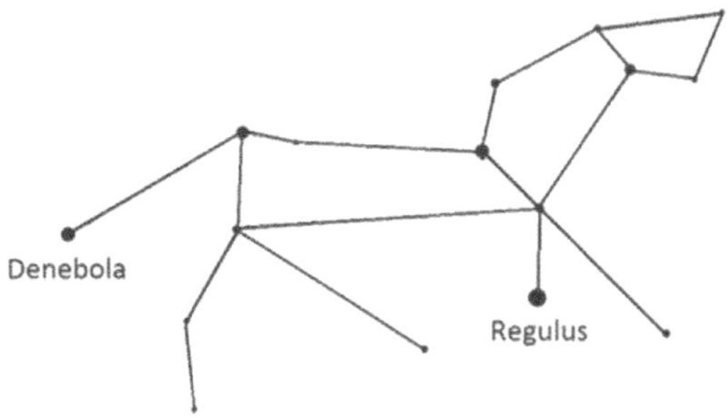

Figure 30. The constellation Leo

The constellation Leo occupies the most space of any of the other constellations. Its brightest star is Regulus. Regulus is a magnitude 1.4 star and is located roughly on the ecliptic. Regulus is a trinary star. Another bright star in the constellation, Denebola, is known as the lion's tail. Denebola is approximately 30° away from Regulus. Algieba and Rasalas, which form the mane and head of the lion, have magnitudes of 2 and 3, respectively. The constellation Leo is easily visible at the vernal point in May in the northern hemisphere.

Finding the Constellation Leo In the Sky: After locating Polaris, you can locate the constellation Leo by following the pointer stars of the constellation Ursa Major.

The Constellation Virgo

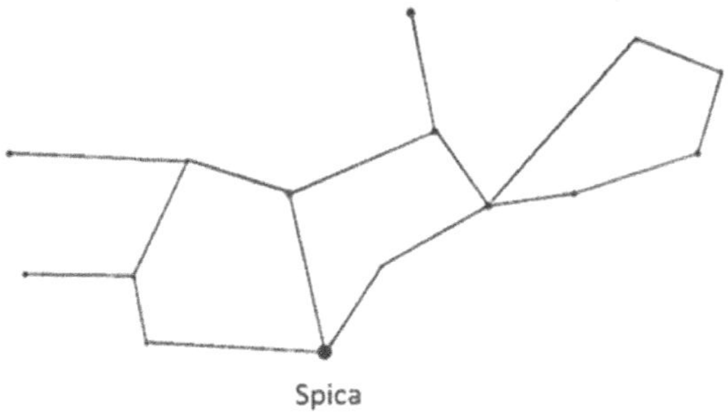

Figure 31. The constellation Virgo

The brightest star in the constellation Virgo is Spica. Spica is the 16th brightest star in the sky. This giant blue star is 260 light-years away from the Earth. It is about twice the size of the Sun and has a brightness of 12,100 times that of the Sun.

There are many galaxies in the constellation Virgo. These galaxies, which are approximately 50,000 light-years away, can only be observed through a telescope. It is hard to tell the nebulae apart from one another when observing through a small telescope. They appear as smudges with varying brightnesses. The galaxy cluster here is called the "Virgo Supercluster." The Virgo Supercluster has 3000 known members.

The constellation Virgo can be observed in spring and summer in the northern hemisphere and in autumn and winter in the southern hemisphere.

The Constellation Libra

There are many binary stars and globular clusters in the constellation Libra. These globular clusters are hard to see with a small telescope. The brightest star in the constellation is the magnitude 2 star Zubeneschamali, located in the northern arm of the constellation. It is 185 light-years away from the Earth. In the southern arm of the constellation is Zubenelgenubi, a binary star of magnitude 3. It is 75 light-years away from the Earth. The constellation Libra can be best observed in the month of June.

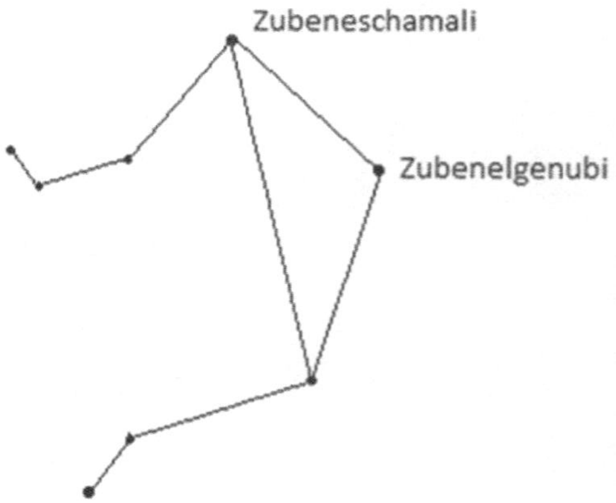

Figure 32. The constellation Libra

The Constellation Scorpio

The brightest star in the constellation Scorpio is Antares, which is also called "the heart of the scorpion." Antares, a binary star, has a diameter 700 times larger than that of the Sun—in other words, the star is larger than the orbit of Mars. It is surrounded by a large, red nebula. Antares's bright light illuminates this nebula. A group made up of Shaula, Sargas, and a few faint stars comprises the scorpion's stinger.

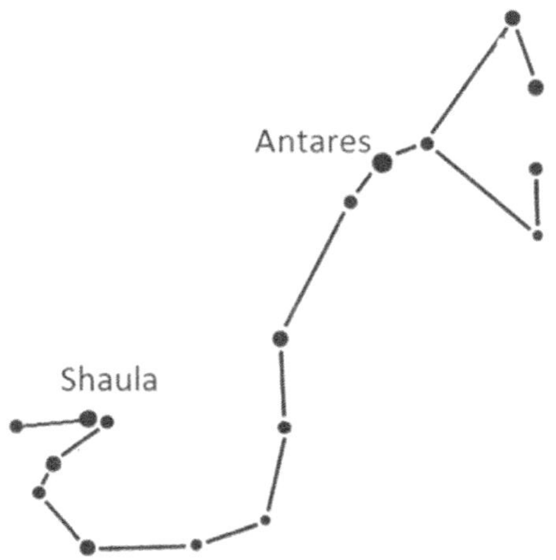

Figure 33. The constellation Scorpio

The one above is a binary star known as Acrab or beta Scorpii. There are 24 galactic clusters in the constellation Scorpio, the largest of which is located near Antares. It is difficult to observe as it is so close to the horizon.

Scorpio stands out among the others not only because of its distinctive shape, but also because it is the brightest constellation in the sky. This constellation is one of the most densely starry sections of the sky. Its shape resembles a scorpion.

Scorpio is relatively easy to spot while observing from a dark location. The majority of the constellation is located in the southern hemisphere. For observers in the northern hemisphere, it can be seen in the month of July.

The Constellation Sagittarius

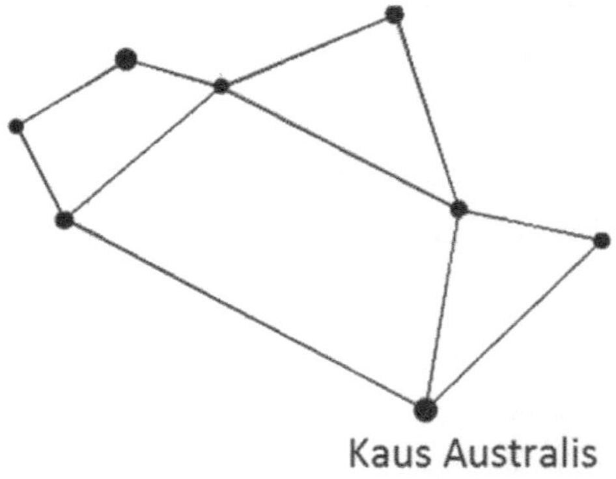

Figure 34. The constellation Sagittarius

The constellation Sagittarius points toward the center of our galaxy, the Milky Way. There is no star in the constellation that particularly stands out in terms of brightness. With its unique shape, Sagittarius is quite easy to recognize. With its various interesting astronomical objects and nebulae, it's worth observing. Its brightest star is Kaus Australis. The giant blue star is 125 light-years away and is 375 times brighter than the Sun.

It can be best observed in August and September in the northern hemisphere, and in spring in the southern hemisphere.

The Constellation Capricorn

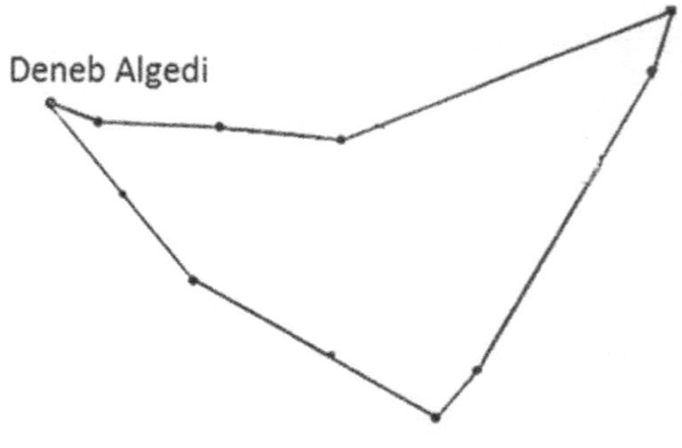

Figure 35. The constellation Capricorn

The constellation Capricorn appears as an upside-down triangle in the sky. Its three brightest stars are Deneb Algedi, with a magnitude of 2.85; Dabih, with a magnitude of 3.05; and Omega Capricorni, with a magnitude of 4.12. Deneb Algedi means "the tail of the goat" in Arabic; it is a binary star that can be seen with the naked eye. It is shown as a single star in very old celestial maps. These two components used to be too close to be distinguished from one another, but over time, they drifted apart. To this day, they continue drifting apart.

The constellation Capricorn can be best observed in the period between July and November in the northern hemisphere and in winter and spring in the southern hemisphere.

The Constellation Aquarius

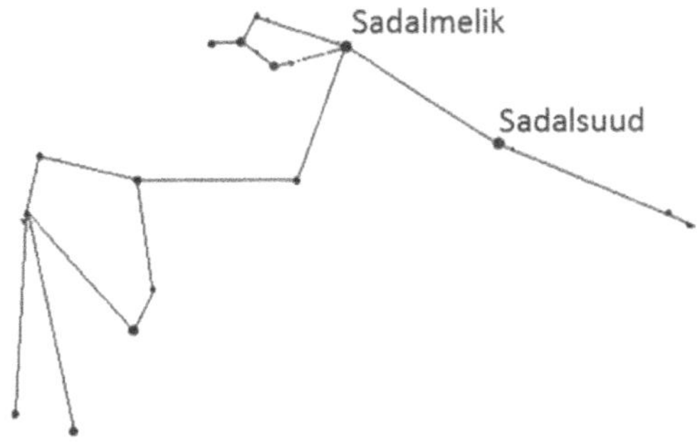

Figure 36. The constellation Aquarius

The constellation Aquarius is located in the southern hemisphere. It is quite far from the Milky Way. There is no particularly bright star in the constellation, and it is hard to observe with the naked eye.

Two important celestial objects can be seen in the constellation with a telescope. One of these is a globular cluster that forms a triangle with the constellation's alpha and beta stars. The brightest star of the constellation is the yellow supergiant beta Aquarii. This star is also known as Sadalsuud. It is 600 light-years away and has a magnitude of 2.9. Alpha Aquarii, also known as Sadalmelik, is a magnitude 2.95 star, and it is 700 light-years away from the Earth. There are many nebulae in the constellation Aquarius. Among these, the Saturn Nebula was first observed by William Herschel in 1782.

The constellation Aquarius can be observed in autumn (particularly in October) in the northern hemisphere and in spring in the southern hemisphere.

The Constellation Pisces

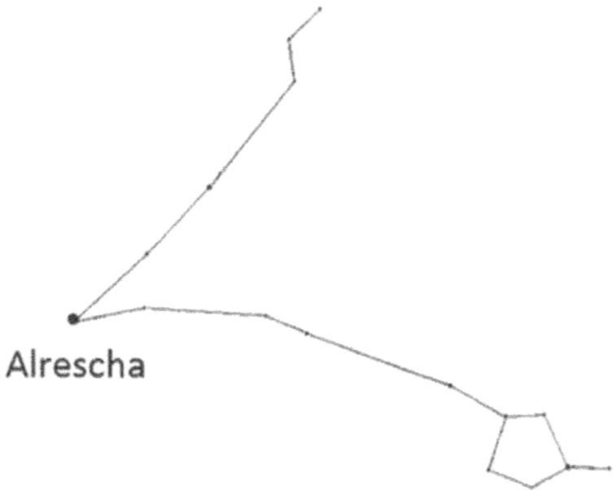

Figure 37. The constellation Pisces

Like the constellation Aquarius, the constellation Pisces is far from the Milky Way. It doesn't have many stars or nebulae. It doesn't have any star clusters or nebulae that are linked to a galaxy that can be observed with a small telescope. With that said, the area covered by the constellation is full of binary stars. There are some galaxies and binary star systems that can provide a beautiful view through novice telescopes.

The constellation appears as a large V shape. At the tip of the V shape, there is Alrescha (Alpha Piscium). This star is 139 light-years away from us, and has a magnitude of 3.62.

Although the constellation is located in the northern hemisphere, it is split into two by the ecliptic. Therefore, the best time to observe it is different for the two hemispheres. It is best observed in autumn in the northern hemisphere and in spring in the southern hemisphere.

XII.
OBSERVATION TOOLS

In the early ages, our ancestors had arranged their lives based on the observations that they made with the naked eye. However, since celestial bodies were millions of light-years away, it was difficult for them to examine them closely. So, mankind has developed many observation tools throughout history. The first of these was the gnomon. Used by the Egyptians, this tool utilized the shadow of a stick erected perpendicularly to the ground. In the present day, we have brought the sky much closer thanks to sophisticated observatories.

Using observation tools increases the quality of our observations. Different observation tools are used for different purposes. Although the term "observation tool" brings telescopes to mind, one of the most important observation tools is a pair of binoculars. Where you will be carrying out your observation, whether you will stabilize your observation tool or not, and many other such choices will affect your choice of observation tool. Needless to say, in order to make this choice, we must first get to know the observation tools:

1. Binoculars

You can get a small, simple pair of binoculars from stores that sell optical devices. Alongside their lightness and portability, the broad field of view that binoculars provide causes them to be among the most indispensable observational tools. (For instance, you can fit the constellations within the field of view of a pair of binoculars.)

As a matter of fact, many of us have a simple pair of binoculars in our homes, but we rarely ever think of turning those binoculars toward the sky. However, a small pair of binoculars is quite an ideal tool when getting to know the sky. With a simple pair of binoculars, we can see comets, galaxies, star clusters, and even very bright nebulae. Furthermore, we can clearly see Jupiter's four major satellites and motions, as well as hundreds of crater formations on the Moon and binary stars that we would be unable to distinguish with the naked eye. With the naked eye, we can observe celestial bodies up to magnitude 5, but a pair of binoculars allows us to observe celestial bodies up to magnitude 8 or 9.

The large lens on the front side of the binoculars is called the objective, and it collects light. The smaller lens that we look through is called the eyepiece. The light that hits the lens is focused onto the eyepiece via prisms inside the binoculars. Thus, we obtain magnification and an image. When choosing a pair of binoculars, the primary consideration should be their magnification. At this point, the objective lens diameter of the binoculars is also important, because the amount of light gathered is proportional to the square of the lens diameter.

Binoculars are defined by two numbers, such as 6x30, 7x35, 8x56, 10x50, 10x70, 15x80, or 20x80. The first number represents magnification, and the second number represents the diameter of the objective lens in millimeters. For instance, an 8x50 pair of binoculars would magnify the image eight times and have an objective lens diameter of 50 millimeters. On a night with no city lights, about 4000-6000 stars can be seen with the naked eye, whereas this number can go up to 150,000 with a pair of binoculars with a lens diameter of 50 millimeters. This is a really good number!

When observing with binoculars, even a minimal amount of shaking profoundly affects the observation. With binoculars with high magnification, the movements of our hands cause the image to shake as well. Therefore, if you are planning on buying a pair of binoculars, you should opt for a pair with eightfold or tenfold magnification. A novice observer can opt for an 8x50 or 10x50 pair of binoculars. A pair of binoculars with a 50-millimeter aperture can gather 50 times more light than the human eye. If you intend to buy binoculars with higher magnification, make sure to support the binoculars with a tripod. I also recommend that you choose the lightest binoculars possible (one kilogram or less). If you learn to observe the sky with binoculars, which are much less expensive than a telescope, using a telescope in the next phase will be much easier.

2. Telescopes

The first person to use and develop the telescope for astronomical purposes was Galileo. Telescopes not only increase the number of celestial bodies that we can see by thousands-fold, but they also allow us a clearer view of them. Although telescopes are more complicated than binoculars, their optical properties are similar. Both aim to gather more light than the eye can gather.

There are finder scopes, capable of scanning wide ranges, mounted on top of telescopes. Celestial bodies are first found through these finder scopes. After being centered, they can be studied through eyepieces with various magnification powers.

A telescope's power is proportional to the amount of light that it gathers. As the aperture of a telescope increases, its ability to gather light increases as well. A telescope's ability to gather light is determined by the diameter of its objective (mirror or lens). Visual observers usually opt for low magnification (below 50x). Sometimes, a magnification between 50x and 150x is chosen, and sometimes, albeit seldom, a high magnification (above 150x) may be preferred. Telescopes with high magnification are primarily used for planetary observations. Contrary to common belief, the magnification of a telescope has no relation to its costliness, nor does the price of a telescope increase along with its size.

Calculating The Magnifying Power Or Magnification

We can find the magnifying power of a telescope by dividing the focal length of the telescope by the focal length of the eyepiece. Focal length is the distance that beams travel from the lens to the point where they come together in focus, expressed in millimeters. The focal length to aperture ratio is called the fratio. For instance, the fratio of a telescope with a diameter of 100 millimeters and a focal length of 1000 millimeters is 10, and it is expressed as f/10. Another example: if you are using a telescope with a focal length of 1000 millimeters and a 25-millimeter eyepiece (1000÷25=40), the fratio is 40. This means you get 40x magnification.

Increasing the magnification by twofold reduces the brightness by three-fourths and the clarity by a half. You can change the magnifying power of your telescope by changing the eyepiece. Buying a telescope with inter-changeable eyepieces will provide you with this option. If you reduce the focal length of the eyepiece, the telescope's magnifying power will increase. However, infinite magnification is not possible because there is a limit to the amount of light that you can gather. Beginners can start their observations with a telescope with a 60-millimeter aperture. With such a telescope, one can easily observe the rings of Saturn and the phases of Venus.

While buying a telescope, you can also buy a few accessories that are suitable for your purpose, such as eyepieces, a tracking mechanism[68] for astrophotography (photographing the sky), filters, or a mount[69].

Now, let's get to know the different types of telescopes. Telescopes can be divided into three categories: refracting telescopes, reflecting telescopes, and compound telescopes.

1. Refracting Telescopes

In 1609, Galileo made an instrument that magnified an object's appearance by threefold by placing lenses inside a tube. Then he magnified this image 32 times with two devices that he built, thus beginning the first modern celestial observation through a telescope. A while later, Kepler made a telescope that had a broader field of vision than Galileo's. These refracting telescopes are, to this day, the most widely used telescopes.

68 When using a telescope with a tracking mechanism, you can make the telescope turn toward a particular celestial body by inputting its coordinates into a computer.

69 A telescope mount is a piece onto which the optical contraption of the telescope is fitted. A telescope mount allows the telescope to move in specific ways.

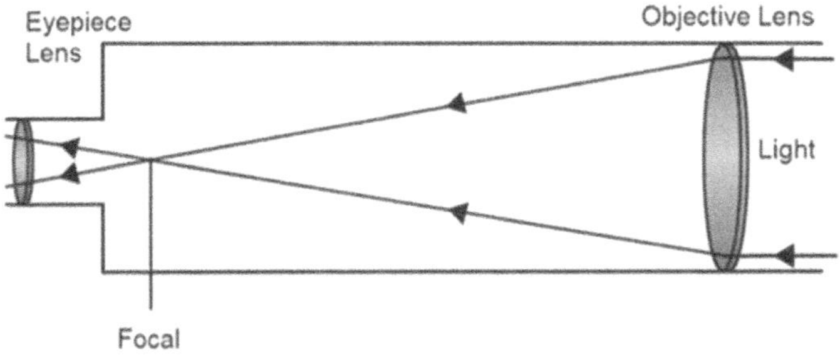

Image 25. Refracting telescope

A refracting telescope has two lenses inside a long tube. The lens at the upper end of the tube gathers the light. The amount of light that this lens gathers determines the amount of detail we can see. At the other end of the tube is a small eyepiece (also called an ocular). The light passes through the large lens and reaches the eyepiece as a small bundle. After this phase, the eyepiece's focus is adjusted.

Refracting telescopes are suitable for observing the Moon, the planets, and binary stars, but they aren't suitable for observing small and faint celestial bodies. The advantages of refracting telescopes are their simple design and ease of use. Their disadvantage is that they are heavy and long.

2. Reflecting Telescopes

In reflecting telescopes, light enters a straight, wide tube and is reflected off of the concave mirror inside the tube. The gathered beams hit a secondary mirror situated at the open end of the tube, after which they arrive at the eyepiece, thus creating an image. Such telescopes usually have spider vanes (fine, X-shaped wires) strung at the open end of the tube to hold the secondary mirror. These spider vanes can be seen through the eyepiece before the focus adjustments are made. However, adjusting the focus should make the vanes invisible.

There are two kinds of reflecting telescopes. The most widely known among these is the Newtonian telescope, devised by and named after Isaac

Newton. Other types of reflecting telescopes are Cassegrain telescopes and compound or catadioptric telescopes.

a. Newtonian Telescopes

Newtonian telescopes have a concave mirror that gathers the light and focuses it onto a flat secondary mirror. This secondary mirror reflects the image to the eyepiece at the opening of the main tube.

These telescopes are usually calculated in inches; 8, 12, 16, 20. One inch corresponds to 25.4 millimeters. Telescopes used in observatory standards are 14 inches, whereas telescopes for novice observers are around 6.8 inches. Newtonian telescopes became one of the most widely used telescopes because of the simplicity of their design and their affordable prices. However, the loss of light caused by the secondary mirror leads to reduced image quality.

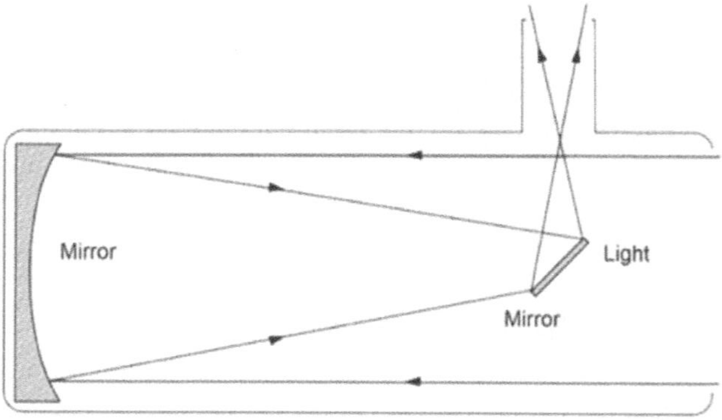

Image 26. Newtonian telescopes

b. Cassegrain Telescopes

These telescopes are called Cassegrain telescopes because they were devised by Laurent Cassegrain in 1672. Cassegrain telescopes consist of a large concave mirror and a convex secondary mirror. First, the beams are gathered by the concave mirror, and then they are reflected onto the secondary mirror. The beams reflecting off of this mirror pass through the hole at the center of the primary mirror and are focused onto the eyepiece. Thus, an image is obtained.

Cassegrain telescopes cost more than Newtonian telescopes. While Cassegrain telescopes are shorter, take up less space, and are easily portable, they require more maintenance because they are more sensitive than other types of telescopes.

3. Compound (Catadioptric) Telescopes

As the name would imply, these kinds of telescopes use both a mirror and a lens, thus combining the advantages of both reflecting and refracting telescopes. There are three kinds of compound telescopes: the Schmidt-Cassegrain, the Maksutov-Cassegrain, and the Schmidt-Newtonian. The Schmidt-Cassegrain is the most widely used type of compound telescope.

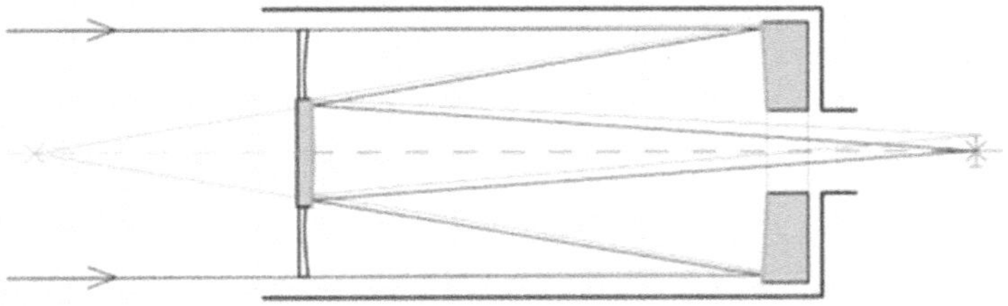

Image 27. Schmidt-Cassegrain type telescopes

In a Schmidt-Cassegrain telescope, the light first passes through a fine correcting lens and hits a concave mirror, which is the primary mirror of the telescope. The beams reflecting off of it hit the concave secondary mirror behind the lens. Then they are reflected off of this mirror as well, after which they pass through the gap at the center of the primary mirror and are focused onto the eyepiece behind the primary mirror. Among various types of telescopes, the Schmidt-Cassegrain is the most modern.

Optically, a compound telescope offers a sharp, clear image within a broad field of view. It's sturdy and low-maintenance. It is, however, more costly, as it contains both a lens system and a mirror system.

XIII.
TIME AND THE CALENDAR

"Time" has remained an important concept through-out the history of humanity. Time was created as a result of the struggle to survive, in order to reduce the unknowns that the changing seasons, the night, and the day may bring.

People of the antique era devised various methods to measure time. The Egyptians made use of the shadow cast by a stick erected perpendicular to the ground. This tool, also known as the gnomon or the sundial, is thought to be the first known observation tool. However, since sundials didn't work at night, water clocks were invented: water dripped out of a container with a hole in the bottom, and time was determined by measuring the water level in the container.

Once the transition to settled life was made and agricultural activities began to be carried out, the notion of "time" became even more important, and the calendar[70] was created. Usually, one of two methods was used: one utilized the periodical motions of the Sun, and the other utilized the periodical motions of the Moon.

The Egyptian calendar consisted of three seasons arranged according to the fluctuations of the Nile: Akhet, also known as the "Season of the Inundation," was between June 21 and October 21; Peret, also known as the "Season of the Emergence," was between October 21 and February 21; and Shemu, which corresponded to the summer, was between February 21 and June 21. Each season consisted of 4 months, and each month consisted of 30 days. The calendar was made up of 10-day weeks that added up to 365 days. The Egyptians knew that one astronomical year had 365 days, so they added a 5-day period to it and created the first official calendar in history.

70 A calendar is the regulation of time according to certain standards.

The Egyptians divided each day into two 12-hour terms, and each day began at sunrise and lasted until the next sunrise. They divided the night into 12 hours based on the motions of the stars rising and setting in the sky. They detected the first star to rise within every 10 day cycle (also known as a "decan") of every year. They established the interval of time between the rising of the first star and the next within a decan as one hour.

Now, let's have a look at other civilizations' notions of time: for the Assyrians, one "week" consisted of 5 days. The Chaldeans divided the week into sevenday intervals and regulated the hours of the day according to the seven celestial bodies found on the zodiac belt. Since their approach toward the planets was based on their distances from the Earth, the following order was established: Saturn, Jupiter, Mars, the Sun, Venus, Mercury, and the Moon. The Chaldeans left us with the names of the days that we still use today: Saturday got its name from Saturn; Sunday, from the Sun; Monday, from the Moon; Tuesday, from Mars; Wednesday, from Mercury; and Thursday, from Venus.

The Sumerians based one year on sexagesimal mathematics, but the lunar calendar also played a part. Lunar calendars were created based on the phases of the Moon: 12 months added up to 354 days. An extra month was added in order to sort out the discrepancy regarding the seasons.

The Greek and Roman calendars were based on the Babylonian calendar. The Greek calendar, like the Babylonian calendar, used to have two seasons: summer and winter. In 650 BC, this system was replaced by the four-season year system.

Romulus, the founder of Rome, developed the first official Roman calendar in 738 BC, which was based on the rotation and phases of the Moon. This calendar began with the month of March, which was named after the Roman god Mars[71]. Every year began in March and ended in December. As a result, the year was divided into ten months (Martius, Aprilis, Maius, Junius, Quintilis, Sextilis, September, October, November, and December). The months of January and February were not included in this calendar because they were months when people were concerned with shelter, warmth, and food—with such concerns, it didn't matter what day it was. Over time, the Roman calendar grew

71 It was believed that the Romans descended from Mars, the god of war. Therefore, they held Mars in high regard—even more so than the other gods.

longer or shorter depending on the political motives of the sovereigns who ascended the throne: the Romans shortened the year to get rid of the politicians they disliked, and they lengthened the year when people they supported were in charge. In order to clear up this confusion, Julius Caesar devised the Julian calendar, named after him, in 46 BC, which was also congruent with the solar year. The Julian calendar established that one year was 365 days and 6 hours long; months with an odd number of days were determined to be 31 days long, while the remaining months were determined to be 30 days long. Furthermore, the month of February was given 29 days instead of 28, and 30 days in leap years.

The Gregorian calendar that we use today is similar to the Julian calendar. One astronomical year corresponds to 365 days, 5 hours, and 49 minutes. The Julian calendar is 11 minutes longer than an astronomical year. This corresponds to 3 days every 400 years. Therefore, the countries that adopted the Gregorian calendar eliminated the error of the Julian calendar by skipping to October 15, 1582, right after October 4, 1582.

Measurement Of Time

There are two methods of time measurement that are based on the Earth's rotation around itself. One is solar time, which is based on the Sun, and the other is sidereal time, which is based on the stars. Now, let's look into the concepts of apparent solar time, mean solar time, and sidereal time.

Sidereal Time

Sidereal time is the interval of time between two consecutive meridian transits of the point of Aries—that is, the vernal equinox. This interval of time is quite close to the interval of time between two consecutive meridian transits of a star. The difference between them is equal to the precession of the point of Aries with respect to the stars.

When measuring sidereal time, we use the hour angle of the first point of Aries instead of just any star. When the first point of Aries is over the meridian of a location, the sidereal time of that location is 0 hours. The hour angle of a star is measured differently from observation spots at different longitudes.

Sidereal Day: Since the motion of the first point of Aries is not regular with respect to the stars, the duration of sidereal days varies, albeit little. Before delving into the sidereal day, we must first understand the concept of a day: one day is the interval of time between two consecutive sunsets. A day, which can also be defined as one full rotation of the Earth around itself, can be measured in two different ways. If we calculate this cycle based on the stars, we will have calculated the sidereal day. In Latin, the word sidereal means "relating to the stars." One sidereal day is the time it takes for the Sun to complete one full rotation with respect to the stars. Since the stars are so far away, the Earth's position changes very little over the course of its orbital motion with respect to the stars. One mean sidereal day corresponds to 23 hours, 56 minutes, and 4.1 seconds.

Sidereal Month: The time interval between two consecutive moonrises with respect to the stars is called a sidereal month. This corresponds to 27.32 days. A sidereal year is the interval of time between the Sun's passing through any fixed star and its return to the same position (as observed from Earth). One sidereal year corresponds to 365 days, 6 hours, 9 minutes, and 9 seconds, or 365.2563 days. Because of precession, one sidereal year is 20 minutes longer than one tropical year.

Solar Time

Solar Day: If we measure the day based on the apparent motion of the Sun, we will have calculated the solar day. An apparent solar day is described as the interval of time between two consecutive transits of the Sun over the observer's meridian. In other words, it equals the interval of time it takes for a point on the surface of Earth and directly across from the Sun to return to its position directly across from the Sun again.

The length of an apparent solar day fluctuates throughout the year. There are two reasons for this: one is that the Sun's yearly apparent motion is not along the celestial equator, through which time is measured, but along the ecliptic. The other is that the Earth does not move at a consistent pace throughout its irregular orbit around the Sun.[72]

Since the apparent solar day is not regular or constant, it is not a convenient measure of time for daily use. Therefore, the "mean solar day" is used. The interval of time between two consecutive transits of the Sun from the observer's meridian is called a mean solar day. The duration of the mean solar day is a constant 24 hours. Our daily lives are regulated according to mean solar time.

What Are Lunar And Solar Years?

The amount of time it takes for the Moon to wax and become full is called a "month." The interval of time between two consecutive new moons is called a "synodic month." A synodic month is approximately 29.5 days long. 12 synodic months comprise one "synodic year." A synodic year is approximately 354 days, 8 hours, 48 minutes, and 36 seconds long.

The regulatory basis of the seasons is the tropical year, meaning the solar year. One solar year equals the interval of time between two consecutive transits of the Sun over the vernal equinox. In other words, it is the amount of time it takes for the Earth to complete one full lap around the Sun. One solar year equals 365 days, 5 hours, 47 minutes, and 48 seconds, or 365,24219 mean solar days.

One lunar year is about 11 days shorter than one solar year. Therefore, in the Gregorian calendar that we use today, the beginning of the lunar calendar shifts forward by 11 days every year.

72 The Earth moves relatively faster when it is closer to the Sun. For instance, for those located in the northern hemisphere of the Earth, the Sun appears to move faster during the winter months. Consequentially, the Sun's apparent motion through the ecliptic becomes irregular.

What Is The Difference Between A Tropical Year And A Sidereal Year?

The tropical year begins with the Sun's transit over the vernal point and is based on the Sun's return to the same point. One tropical year is 365 days, 5 hours, and 48 seconds long. The sidereal year, on the other hand, is based on the time it takes for the Earth to complete one full revolution around the Sun. In other words, it is the interval of time between two consecutive appearances of the Sun in a constellation as seen from the Earth. Since the vernal point isn't fixed due to precession, one sidereal year lasts 20 minutes longer than one tropical year.

Time Zones

There are 24 recognized meridians; they start with the prime meridian at Greenwich[73] and are spaced at equal intervals. 12 of these 24 meridians are east of the prime meridian, and the remaining 12 are west of the prime meridian. The angle between two neighboring standard meridians is 15°. In other words, as the world rotates by 360° in 24 hours, its speed is 15° per hour. (360÷24=15°)

The area between 7°30′E and 7°30′W of Greenwich is the zero standard[74] time zone. Every 15° zone corresponds to an hour of time. Places located in the same time zone use the same mean solar time. This is called "zone time."

73 The Greenwich meridian is the meridian passing through the Royal Observatory in Greenwich. It is accepted as the prime meridian (0°). (Greenwich is a suburb in southeast London.)

74 The complications caused by solar time have led to the emergence of the concept of standard time. Around the end of the 19th century, iron transportation became prevalent, making it necessary to establish a standard time.

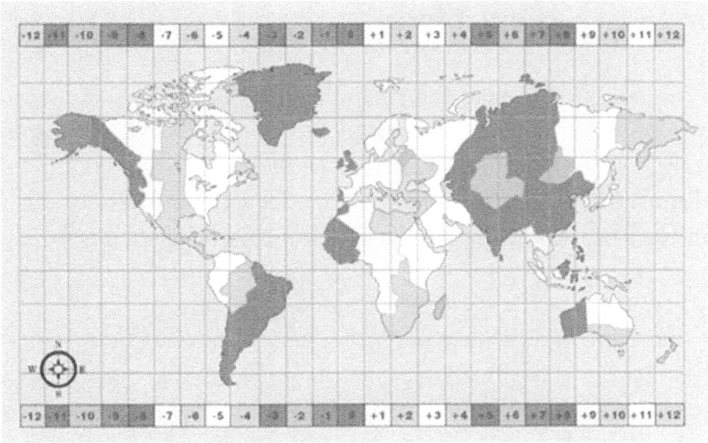

Image 28. Map of the timezones of the world

What Is Greenwich Mean Time (GMT)?

GMT is the mean time at Greenwich. This term is used for the zone time defined by the Greenwich meridian, which is the international prime meridian. It is expressed as (UT) or (GMT). Universal time (UT) is the GMT, expressed in accordance with the 24-hour time standard. For instance, 1:00 p.m. in GMT is 13:00 in UT.

Every country uses the mean solar time—that is, the zone time—of the longitude that passes through it or is closest to it. Countries such as China, Russia, and the United States are spread out over vast lands, thus necessitating the use of multiple zone times. Standard time is the time in a region when daylight saving time or other similar practices are not in use. For instance, the standard time in Turkey is UT+2 throughout the year, across all meridians. This means that Turkey uses 2 hours eastern zone time with respect to Greenwich. Turkey lies 2 hours east of the prime meridian. When it is 10:00 in Turkey, the zone time in Greenwich is 08:00. As we move eastward from Greenwich, time advances (+), and as we move westward, time is set back (–).

Daylight Saving Time

Daylight saving time is the periodical practice of adjusting clocks by a certain amount in order to make less use of daylight in the morning and more use of it in the afternoon. Modern daylight saving time was first proposed in 1895 by a New Zealander entomologist called George Vernon Hudson. After that day, many countries adopted and began to use this practice.

Daylight saving time practices may vary from one country to the next. Usually, however, countries that use daylight saving time set the clocks forward by an hour at the beginning of spring, and set them back by an hour in autumn.

XIV.
CASTING A NATAL CHART

The days of hand-made natal charts are long gone. Thanks to the astrology programs of the present day, it is quite simple to cast natal charts. Although casting natal charts is arduous and time-consuming, it serves a crucial purpose for the astrologist: the planet and house placements that gradually take shape throughout the creation of the natal chart prepare the astrologist to interpret the natal chart.

Most people believe that casting a natal chart requires a lot of complex calculations. Indeed, the process may initially give off this impression. However, with regular practice, you will notice that you can cast natal charts with more ease.

Here is what you need to create a natal chart:

1. *Birth information*
The day, month, year, and time of your birth. Some astrologists believe that the time of birth is when the umbilical chord is severed, while others believe it is when the child draws his first breath.

2. *Ephemeris containing the year of birth*
An ephemeris is a celestial journal containing planetary positions and astronomical data. *Raphael's Ephemeris* is one of the most well-known and reliable ephemerides.[75] *Raphael's Ephemeris* is a noon ephemeris, so the positions it provides are listed based on the noon of a specific day. In *Raphael's Ephemeris*, which has been published every year since 1860, Pluto positions prior to 1934 are not listed.

75 Other ephemerides that you can use include *The American Ephemeris For The 20th Century:* 1900 To 2000 by Neil F. Michelsen, California; *Europa Ephemeride 1900-1950* and *Europa Ephemeride 1950-2005* by Dr. Wolf Metz Import, Zurich; and *Rosenkreuzer Ephemeride 1900-2000* by Bauer Publishing, Freiburg.

For Pluto positions before 1934, I recommend *The Influence of the Planet Pluto*, written by Elbert Benjamine and published by Aries Press. Some ephemerides show midnight placements—that is, the placements listed are for 0 hours—indicating the commencement of the day. We will be using a midnight ephemeris for our calculations.

3. *Tables of houses*
We will be using these to calculate the starting degrees of the ASC and the houses. Usually, tables that have been prepared in accordance with the Placidus house system are used. There are also tables of houses prepared in accordance with other house systems.[76]

4. **Latitude and longitude of the place of birth**
You can find more detailed information on the latitude and longitude of a place of birth in geographical atlases.

5. *Information on standard time and daylight saving time*[77]

6. *A calculator*

7. *An empty chart (There are two empty charts among the last pages of this book.)*

Creating A Natal Chart, Step By Step

Step 1: First, we convert the time of birth to Greenwich Mean Time (GMT).

First, we convert the time of birth to 24-hour time format. Let's say, for instance, that the individual in question was born at 5:00 p.m. (a.m. means before noon and p.m. means after noon.) According to the 24-hour time format, the individual's time of birth will be 17:00. What we

76 For Placidus house tables: *The Michelsen Book Of Tables* by Neil F. Michelsen, published by ACS Publication and *Die globalen Häusertabellen*, published by Otto Wilhelm Bart, Munich.

77 You can find more information on daylight saving time practices in Turkey at www.astroturkiye.com/YazSaatleri.html

are doing here is adding 12 to the value in p.m. Had this individual been born at 5:00 a.m., then we would take the time of birth as 5:00 without any further calculations.

Next, we find the time difference from an atlas[78]. If daylight saving time or war time was in practice during the relevant period, we must take that into consideration as well. The value discovered for the time difference is either added to or subtracted from the time of birth, which is in the 24-hour time format. If the place of birth is west of Greenwich, we add; and if it is east of Greenwich, we subtract.

Sometimes, the time we obtain may be greater than 24 hours. In such cases, we subtract the result from 24 (as in 24 hours) and use the information listed in the ephemeris for the day after the day of birth. The result we get will be used to calculate planetary placements. Let's put the first step to the test with an example.

Example

Name: Al Pacino
Date Of Birth: April 25, 1940
Place Of Birth: Manhattan, New York
Time Of Birth: 11:02 a.m.
Latitude: 40 deg 46 min North
Longitude: 73 deg 59 min West
Time Difference: EST 5:00

Calculating Greenwich Mean Time (GMT)

78 Thomas G. Shanks's *The American Atlas* and *The International Atlas* can be used for this.

	Hour	Minute	Second
Time Of Birth	11 h	02 min	00 sec
Time Zone (73° West) +	05 h	00 min	00 sec
GMT At The Time Of Birth	16 h	02 min	00 sec

UT (Universal Time) is GMT, expressed according to the 24-hour time format. In that case, UT is 16:02:00.

Step 2: We calculate the Local Sidereal Time (LST) of the birth.

Local Sidereal Time (LST) is the local sidereal time used for various meridians. For this, we need the sidereal time at Greenwich (GST). The GST, or Greenwich Sidereal Time, is the local sidereal time of Greenwich.

i. First, let's calculate the GST, or the Greenwich Sidereal Time.

In order to calculate the Greenwich Sidereal Time (GST), we convert GMT to sidereal time. One sidereal day is 23 hours, 56 minutes, and 4,09 seconds long. The stars rise and set 3 minutes and 56 seconds earlier than in mean solar time. This corresponds to 9,865 seconds per hour.

You may recall that in the example above, we discovered that the GMT at the time of birth was 16:02:00. What we do now is convert GMT to sidereal time. For this, we find the Mean Time correction value (for the sake of simplicity, we will refer to the Mean Time correction value as *the correction value* from this point on). Again, our GST was 16 hours, 02 minutes, and 00 seconds. A correction at the rate of 9,865 seconds was required per hour. Based on this information;

The correction value for 16 hours; $16 \times 9{,}865 = 157.84$ seconds

The correction value for 2 minutes; $2/60 \times 9{,}865 = 0.3288$ seconds

The total correction value for 16 hours and 2 minutes;

= 157,84 seconds + 0,3288 seconds

= 158,17 seconds

= 2.63 minutes

= 2 minutes and 38 seconds

We also need the sidereal time for the UT day. We can find this value in the row relevant to the birthday in the ephemeris. (The UT values stated in midnight ephemerides are for midnight.)

MARCH 1940 — **00:00 UT**

Day	Sid.t	☉	☽	☿	♀	♂	♃	♄	♅	♆	♇	☊	☊	⚸	⚷
F 1	10 34 35	10♓15'48	8♐57	28♓10	21♈ 5	8♉44	11♈57	27♈54	18♉29	24°R32	1°R 0	20°R50	22♎23	18♓41	14°R10
S 2	10 38 32	11°16'00	22° 0	28°53	22°15	9°25	12°11	28° 0	18°31	24♍30	0♌59	20♎49	22°20	18°47	14♋ 9
S 3	10 42 28	12°16'10	4♑43	29°27	23°25	10° 5	12°24	28° 6	18°33	24°28	0°58	20°47	22°16	18°54	14° 8
M 4	10 46 25	13°16'19	17°11	29°52	24°35	10°46	12°38	28°12	18°34	24°27	0°57	20°43	22°13	19° 1	14° 6
T 5	10 50 21	14°16'26	29°26	0♈ 8	25°44	11°26	12°52	28°18	18°36	24°25	0°56	20°37	22°10	19° 7	14° 5
W 6	10 54 18	15°16'31	11♒31	0°R15	26°54	12° 7	13° 5	28°24	18°38	24°24	0°55	20°31	22° 7	19°14	14° 4
T 7	10 58 15	16°16'35	23°30	0°12	28° 3	12°47	13°19	28°31	18°40	24°22	0°54	20°24	22° 4	19°21	14° 3
F 8	11 2 11	17°16'37	5♓23	0° 1	29°13	13°27	13°33	28°37	18°42	24°20	0°53	20°18	22° 0	19°27	14° 2
S 9	11 6 8	18°16'37	17°12	29♓41	0♉22	14° 8	13°46	28°43	18°44	24°19	0°52	20°13	21°57	19°34	14° 1
S 10	11 10 4	19°16'35	29° 1	29°13	1°31	14°48	14° 0	28°50	18°46	24°17	0°51	20° 9	21°54	19°41	14° 0
M11	11 14 1	20°16'31	10♈49	28°38	2°40	15°28	14°14	28°56	18°48	24°15	0°51	20° 8	21°51	19°47	13°59
T 12	11 17 57	21°16'25	22°41	27°56	3°48	16° 8	14°28	29° 3	18°51	24°14	0°50	20°D 7	21°48	19°54	13°59
W13	11 21 54	22°16'17	4♉38	27° 9	4°57	16°49	14°42	29°10	18°53	24°12	0°49	20° 8	21°45	20° 1	13°58
T 14	11 25 50	23°16'07	16°43	26°18	6° 5	17°29	14°56	29°16	18°55	24°10	0°48	20°10	21°41	20° 7	13°58
F 15	11 29 47	24°15'55	29° 2	25°23	7°13	18° 9	15°10	29°23	18°57	24° 9	0°48	20°11	21°38	20°14	13°57
S 16	11 33 43	25°15'41	11♊36	24°27	8°21	18°49	15°24	29°30	19° 0	24° 7	0°47	20°13	21°35	20°21	13°57
S 17	11 37 40	26°15'24	24°32	23°30	9°29	19°30	15°38	29°37	19° 2	24° 5	0°46	20°R13	21°32	20°27	13°57
M18	11 41 37	27°15'05	7♋51	22°34	10°37	20°10	15°52	29°43	19° 5	24° 4	0°46	20°13	21°29	20°34	13°D57
T 19	11 45 33	28°14'44	21°37	21°40	11°44	20°50	16° 7	29°50	19° 7	24° 2	0°45	20°12	21°26	20°41	13°57
W20	11 49 30	29°14'20	5♌50	20°49	12°51	21°30	16°21	29°57	19°10	24° 0	0°44	20° 9	21°22	20°47	13°57
T 21	11 53 26	0♈13'55	20°28	20° 1	13°59	22°10	16°35	0♉ 4	19°12	23°59	0°44	20° 7	21°19	20°54	13°57
F 22	11 57 23	1°13'26	5♍26	19°18	15° 5	22°50	16°49	0°11	19°15	23°57	0°43	20° 4	21°16	21° 1	13°58
S 23	12 1 19	2°12'56	20°37	18°40	16°12	23°30	17° 3	0°18	19°17	23°55	0°43	20° 2	21°13	21° 7	13°58
S 24	12 5 16	3°12'24	5♎50	18° 7	17°18	24°10	17°18	0°25	19°20	23°54	0°42	20° 1	21°10	21°14	13°59
M25	12 9 12	4°11'49	20°57	17°40	18°24	24°50	17°32	0°32	19°22	23°52	0°42	20°D 0	21° 6	21°21	13°59
T 26	12 13 9	5°11'13	5♏48	17°19	19°30	25°30	17°46	0°40	19°25	23°50	0°41	20° 1	21° 3	21°27	14° 0
W27	12 17 6	6°10'35	20°17	17° 3	20°36	26°10	18° 1	0°47	19°28	23°49	0°41	20° 2	21° 0	21°34	14° 1
T 28	12 21 2	7° 9'55	4♐20	16°54	21°41	26°50	18°15	0°54	19°31	23°47	0°40	20° 3	20°57	21°41	14° 2
F 29	12 24 59	8° 9'13	17°56	16°D50	22°47	27°30	18°29	1° 1	19°33	23°46	0°40	20° 4	20°54	21°47	14° 3
S 30	12 28 55	9° 8'29	1♑ 5	16°52	23°51	28° 9	18°44	1° 8	19°36	23°44	0°39	20° 4	20°51	21°54	14° 4
S 31	12 32 52	10♈ 7'44	13♑52	17♓ 0	24♉56	28♉49	18♈58	1♉16	19♉39	23♍42	0♌39	20°R 4	20♎47	22♓ 1	14♋ 5

Image 29. Midnight ephemeris for April 1940

Finally, we add up these three values.

	Hour	Minute	Second
GMT of birth	16 h	02 min	00 sec
Correction value		02 min	38 sec
Sidereal time for the UT day +	14 h	11 min	26 sec
Greenwich Sidereal Time (GST)	30 h	15 min	04 sec

ii. Now, let's calculate the LST, or the Local Sidereal Time.

The LST, along with the tables of houses, is used for finding house cusps. Each meridian has a different LST. We'll need some general information to find the LST. Let's refresh our memory:

- One hour equals 15° of longitude.
- The Earth completes one full rotation around itself in 24 hours, which means that it rotates 360° in 24 hours. (360/24 = 15°)
- 1° of longitude equals 4 minutes.
- 60 minutes equals one hour.
- 60 seconds equals one minute.

First, we convert the longitude degree of the place of birth to hours. For this, we begin by converting the longitude degree to minutes. Then we divide this by 15 to obtain this value in hours. (It is divided by 15 because 1 hour of time equals 15° of longitude.) We then express this value (which is in hours) in hours, minutes, and seconds. If the resulting value is west of Greenwich, we subtract it from the GST value. If it is east of Greenwich, we add it to the GST value.

Another consideration is whether the birthplace is in the northern or southern hemisphere. If it is in the southern hemisphere, we proceed in the same manner, but if the value we obtain is less than 12, we add 12 to it, and if it is greater than 12, we subtract 12 from it. We take not the signs that we obtained, but those right across from them.

Example

Longitude of the place of birth: 73 deg 59 min West

Longitude value in hours: = (73 + 59/ 60)° /15

= 4,9322 hours

= 4 h 55 min 56 sec

	Hour	Minute	Second
Greenwich Sidereal Time (GST)	30 h	16 min	04 sec
Longitude of the place of birth –	4 h	55 min	56 sec
Total (LST)	25 h	20 min	8 sec

Since the resulting LST value is greater than 24 hours, we subtract 24 from it.

	Hour	Minute	Second
Total (LST)	25 h	20 min	08 sec
–	24 h	00 min	00 sec
LST	01 h	20 min	08 sec

The LST value we obtain will be used for calculating house cusps.

Step 3: We calculate the planets' placements.

We calculate the planets' placements according to Greenwich time at the time of birth. This is because the ephemeris is based on Greenwich. First, we find the GMT of the birthplace in hours or minutes. Then we find out what fraction of a day this value corresponds to. For this, we divide the value in hours by 24 (1 day equals 24 hours) or the value in minutes by 1440 (1 day equals 1440 minutes). The constant value that we obtain will be used for calculating the planets.

Then we find the distance traveled in one day by the planet we wish to calculate. We subtract the longitude degree of the individual's day of birth from the longitude degree of the day after the day of birth. If the planet is in retrograde, we subtract the planet's longitude degree on the day of birth (UT) from the longitude degree of the day after. Then, we multiply this difference by the constant value that we previously calculated and add the resulting value to the position on the day of birth. With slow-moving planets, the constant value does not need to be taken into consideration.

Example

Calculating The Placement Of The Sun

		Hour	Minute	Second
Sun placement for April 26, 1940		05 h	36 min	25 sec
Sun placement for April 25, 1940	−	04 h	38 min	02 sec
Difference		00 h	58 min	23 sec

= 58 min + 23/60 min

= 58,383 min

We multiply the number we obtained by the constant value to determine how far the Sun has moved after midnight.

Finding the Constant Value:

The UT (GMT) was 16:02:00. We can find the constant value by converting this into minutes and dividing it by 1440.

Constant value = (16 × 60 + 2) /1440

Constant value = 0,668 min

Then we multiply the difference in minutes by the constant value that we obtained.

$$= 58{,}383 \text{ min} \times 0.668$$

$$= 38{,}999 \text{ min}$$

	Hour	Minute	Second
Sun placement for April 25, 1940	04 h	38 min	02 sec
+	00 h	39 min	00 sec
Sun placement for April 25, 1940	05 h	17 min	02 sec

Sun placement: 05 deg 17 min 02 sec Taurus

Calculating the placement of the Moon

	Hour	Minute	Second
Moon placement for April 26, 1940	26 h	10 min	47 sec
Moon placement for April 25, 1940	−12 h	27min	56 sec
Difference	13 h	42 min	51 sec

$$= (13 \times 60) \text{ min} + 42 \text{ min} + (51/60) \text{ min}$$
$$= 822{,}85 \text{ min} \times 0{,}668$$
$$= 549{,}664 \text{ min}$$
$$= 09 \text{ deg } 09 \text{ min } 40 \text{ sec}$$

Let's express this value in degrees, minutes, and seconds. To convert a value in minutes to degrees, minutes, and seconds, we first divide the value in minutes by 60. 38,999/60 = 0,649°. This equals the sum of 0° and 0,65°. Now, let's convert 0,65° to minutes. We do this by multiplying 0.65° by 60 (0.65 × 60 = 38.999 min). The 38,999 minutes we obtain equals the sum of 38 minutes and 0,999 minutes. To convert 0.999

minutes to seconds, we multiply it by 60 (0,999 × 60 = 59,94; about 1 second).

The result we obtain is 0 degrees, 38 minutes, and 01 seconds. We add this value to the planetary values on the individual's birthday to determine the placement of the relevant planet.

	Hour	Minute	Second
Moon placement for April 25, 1940	12 h	27 min	56 sec
+	09 h	09 min	40 sec
	21 h	37 min	36 sec

Moon placement: 21 deg 37 min 36 sec Sagittarius

Calculating the placement of Mercury

	Hour	Minute
Mercury placement for April 26, 1940	11 h	52,4 min
Mercury placement for April 25, 1940 −	10 h	23,6 min
Difference	01 h	28,8 min

= (1 × 60) + 28,8 min
= 88,8 min × 0,668
= 59,32 min

	Hour	Minute
Mercury placement for April 25, 1940	10 h	23,6 min
+	59,32 min	
	10 h	82,92 min

Mercury placement: 11 deg 22 min 55 sec Aries

Calculating the placement of Venus

		Hour	Minute
Venus placement for April 26, 1940		20 h	56,4 min
Venus placement for April 25, 1940	−	20 h	2,1 min
Difference		00 h	54,3 min

= 54,3 min
= 54,3 min × 0,668
= 36,27 min

		Hour	Minute
Venus placement for April 25, 1940		20 h	2,1 min
	+		36,27 min
		20 h	38,38 min

Venus placement: 20 deg 38 min 22 sec Gemini

Calculating the placement of Mars

		Hour	Minute
Mars placement for April 26, 1940		15 h	56,9 min
Mars placement for April 25, 1940	−	15 h	17,6 min
Difference		00 h	39,3 min

= 39,3 min
= 39,3 min × 0.668
= 26,25 min

Mars placement for April 25, 1940	15 h	17,6 min
	26,25 min	
	15 h	43,85 min

Mars placement: 15 deg 43 min 51 sec Gemini

Calculating the placement of Jupiter

	Hour	Minute
Jupiter placement for April 26, 1940	25 h	13,8 min
Jupiter placement for April 25, 1940	−24 h	59,4 min
Difference	00 h	14,4 min

= 14,4 min
= 14,4 min × 0,668
= 9,62 min

Jupiter placement for April 25, 1940	24 h	59,4 min
+		9,62 min
	24 h	69,02 min

Jupiter placement: 25 deg 9 min 1 sec Aries

Calculating the placement of Saturn

	Hour	Minute
Saturn placement for April 26, 1940	4 h	32,2 min
Saturn placement for April 25, 1940 –	4 h	24,5 min
Difference	00 h	7,7 min

= 7,7 min
= 7,7 min × 0,668
= 5,14 min

Saturn placement for April 25, 1940	4 h	24,5 min
	+	5,14 min
	4 h	29,644 min

Saturn placement: 4 deg 29 min 39 sec Taurus

Calculating the placement of Uranus

	Hour	Minute
Uranus placement for April 26, 1940	21 h	1,2 min
Uranus placement for April 25, 1940 –	20 h	57,9 min
Difference	00 h	3,3 min

= 3,3 min
= 3,3 min × 0,668
= 2,20 min

Uranus placement for April 25, 1940	20 h	57,9 min
	+	2,20 min
	20 h	60,105 min

Uranus placement: 21 deg 0 min 6 sec Taurus

Calculating the placement of Neptune

	Hour	Minute
Neptune placement for April 25, 1940	23 h	7,5 min
Neptune placement for April 26, 1940	23 h	6,3 min
Difference	00 h	1,2 min

= 1,2 min × 0,668
= 0,8 min

	Hour	Minute
Neptune placement for April 25, 1940	23 h	7,5 min
The value is subtracted as Neptune is retrograde		−0,8 min
Difference	23 h	6,7 min

Neptune placement: 23 deg 6 min 42 sec Virgo Retrograde

Calculating the placement of Pluto

		Hour	Minute
Pluto placement for April 26, 1940		0 h	39,0 min
Pluto placement for April 25, 1940	–	0 h	38,6 min
Difference		00 h	0,4 min

= 0,4 min
= 0,4 min × 0,668
= 0,27 min

		Hour	Minute
Pluto placement for April 25, 1940		00 h	38,6 min
	+	0,27 min	
		38,87 min	

Pluto placement: 00 deg 38 min 52 sec Leo

Calculating the placement of Lunar Nodes

		Hour	Minute
Lunar node placement for April 26, 1940		19 h	57,9 min
Lunar node placement for April 25, 1940	–	19 h	55,9 min
Difference		00 h	02 min

= 2 min
= 2 min × 0,668
= 1,34 min

Lunar node placement for April 25, 1940	19 h	57,9 min
	–	1,34 min
	19 h	56,56 min

True lunar node placement: 19 deg 56 min 33 sec Libra-Aries

Step 4: We must calculate the houses' cusps.

i. Calculating the ASC and MC

As previously mentioned, we will be using the LST to find the house cusps. LST values are given in the upper left corner of the house of tables. The value in the middle of the row indicates the MC. The exact LST value calculated for the individual's place of birth may not be present in the house of tables. In that case, we choose the closest value in the table. For instance, if the LST is 15 h 11 min 48 sec, we choose either the 15 h 08 min 00 sec table or the 15 h 12 min 00 sec table. Among these two values, the second one is closest to our example.

We will need two constant values to find the house cusps: one is the sidereal factor, which will be used for the MC and the other houses, and the other is the latitude factor, which will be used for houses other than those of the MC.

Finding The Sidereal Factor:

We use the table just above and the table just below the value closest to the local sidereal time (LST) of the birthplace. We subtract the LSTs of these two tables from one another. On the other side, the LST of the birthplace is subtracted from the closest ST value. When we divide these two differences by each other, we obtain the sidereal factor.

Example

You may recall that we had determined the LST to be 01 deg 20 min 8 sec.

The closest LST value in the house of tables is 01 deg 20 min 00 sec.

The difference between these two values:

Example LST	01 deg	20 min	08 sec
Closest LST	01 deg	20 min	00 sec
	00 deg	00 min	08 sec

We will find the sidereal factor by converting the value that we obtain to minutes, and then dividing it by four.

The sidereal factor = (8/60) /4 = 0,13/4 = 0,033

Finding The Latitude Factor:

We find the latitude factor by dividing the minute value of the latitude of the individual's birth by 60.

Latitude 40 deg 46 min North

The latitude factor = 46/60 = 0,766

Calculating The MC:

We find the difference between the MC degrees in the relevant LST tables. We multiply the value we obtained by the sidereal factor. We then add the result to the previous MC value.

Example

The LST value was 01 h 20 min 08 sec. We then find the difference between the MC values in the two LST tables closest to this value.

MC value for LST	1 h	24 min	0 sec	22 deg	42 min	Aries
MC value for LST	1 h	20 min	0 sec	21 deg	38 min	Aries
Difference				1 deg	4 min	

We multiply the difference we obtained by the sidereal factor.

$$1 \text{ deg } 4 \text{ min} = (1\times60) + 4$$
$$= 64 \text{ min} \times 0{,}033$$
$$= 2{,}112 \text{ min}$$
$$= 2 \text{ min } 7 \text{ sec}$$

We add this value to the previous MC value.

MC value for LST	1 h	20 min	0 sec	-	21 deg 38 min
	+				2 min 7 sec
					21 deg 40 min 7 sec Aries

MC placement: 21 deg 40 min 7 sec Aries

Right across from the placement of MC is the placement of IC:

IC placement: 21 deg 40 min 7 sec Libra

Calculating the cusps of the ASC and the other houses:

The cusps of the ASC and the other houses differ in every latitude. Therefore, the latitude factor must be taken into consideration. (We didn't use the latitude factor when calculating the MC, because the MC value is the same across all latitudes on the same longitude.)

Finding the ASC:

We multiply the difference between the ASCs in the ST tables just before and just after the latitude of the individual's birthplace by the ST factor. We add the result to the ASC value at the relevant latitude. We find the ASC difference between the relevant latitude and the next latitude from the sidereal table just before the LST of birth. We multiply the result by the latitude factor. The two values we have obtained are either subtracted from or added to one another.

Example

The value in the ASC column of the 40th latitude in the next ST table: 05 deg 22 min Leo

The value in the ASC column of the 40th latitude in the previous ST table: 04 deg 34 min Leo

The difference between the two: 48 min

We multiply the difference by the sidereal factor.

48 min × 0,033 = 1,5984 min = 1 min 36 sec

We add this to the value in the ASC column of the 40th latitude in the previous ST table.

The value in the ASC column of the 40th latitude in the previous table: 04 deg 34 min Leo + 01 min 36 sec = 04 deg 35 min 36 sec

Next, we must find the difference between the ASC columns of the 41st and 40th latitudes.

		Degree	Min
41st latitude	ASC	05	04
40th latitude	ASC	04	34
Difference			30 min

We multiply the difference by the latitude factor.

= 30 x 0,7666
= 22,998 min

Finally, we subtract the value we found according to the latitude factor from the value we found according to the longitude factors.

04 deg	35 min	36 sec	Leo
	22 min	60 sec	Leo
04 deg	57 min	96 sec	

ASC placement: 04 deg 58 min 36 sec Leo

DSC placement: 04 deg 58 min 36 sec Aquarius

Calculating the other house cusps:

We are going to calculate the cusps of the 2nd and 8th houses. The cusps of the other houses are also calculated in the same manner.

Calculating the cusps of the 2nd and 8th house:

The 2nd house cusp of the 40th latitude in the next ST table 26 deg 06 min Leo

The 2nd house cusp of the 40th latitude in the previous ST table 25 deg 14 min Leo

Difference = 52 min

We multiply the difference by the sidereal factor.

$$= 52 \text{ min} \times 0{,}0333$$
$$= 1{,}7316 \text{ min}$$
$$= 1 \text{ min } 44 \text{ sec}$$

We add this to the 2nd house value of the 40th latitude in the previous ST table.

25 deg 14 min Leo + 01 min 44 sec = 25 deg 15 min 44 sec

Then we find the difference between the 2nd house values for the 41st latitude and the 40th latitude.

41. latitude	2nd house	25 deg 28 min
−40. latitude	2nd house	25 deg 14 min

Difference = 30 min

We multiply the difference by the latitude factor.

= 30 x 0,7666
= 22,998 min = 23 min

Then we add the value we found according to the longitude factors to the value we found according to the latitude factor.

25 deg	15 min	44 sec
	23 min	00 sec
25 deg	38 min	44 sec

Cusp of the 2nd house: 25 deg 38 min 44 sec Leo

Cusp of the 8th house: 25 deg 38 min 77 sec Aquarius

GLOSSARY OF ASTRONOMICAL TERMS

Almagest: A book on mathematics, geography, and other fields, written by Ptolemy in the 2nd century AD.

Andromeda: The closest galaxy to the Milky Way.

Aphelion: The point where a celestial body is furthest from the Sun.

Apogee: The point where a celestial body is furthest from the Earth.

Asteroid: The name given to rocks found in the asteroid belt between Mars and Jupiter. Their diameters can reach 300 kilometers.

Astrolabe: A mechanical instrument used to measure angular distances and determine the altitude of the Sun or other celestial bodies.

Astronomical Unit (AU): The unit of measurement of the mean distance between the Earth and the Sun. 1 astronomical unit corresponds to 149.6 million kilometers. Astronomical units are used to express the distances between celestial bodies in the universe.

Astronomical Horizon: Astronomical horizon and true horizon roughly have the same meaning. The true horizon is in line with your eye, whereas the astronomical horizon is perpendicular to your zenith. Because of the massiveness of the Earth, the true horizon and the astronomical horizon will be in the same direction in the observer's location.

Astronomy: The field of science concerned with the motions of celestial bodies, their structures, and their conditions in space.

Aurora: Charged particles of solar origin that enter through the openings in the Earth's magnetic field lines. The Earth's atmosphere causes the gases to fluoresce, which results in a colorful scene.

Azimuth: Azimuth is the number of degrees from the north point to the great circle passing through the star and the zenith, measured clockwise (eastward) along the horizon circle.

Binary Star System: A system made up of two stars orbiting each other.

Claudius Ptolemy/Claudius Ptolemaeus: A mathematician, astronomer, astrologist, and geographer from the antique period.

Celestial Equator: The virtual great circle passing at a distance of 90° from both the northern and southern celestial poles.

Comet: Celestial bodies made up of ice and gases that follow a mostly parabolic orbit around the Sun. Comets evaporate as they get closer to the Sun, creating a tail that extends away from the Sun. They get their names from the fact that they reflect sunlight.

Conjunction: When two celestial bodies can be observed together in the sky, we say they are in conjunction.

Constellation: Stars that appear as a group and form distinct shapes in the sky based on the direction from which we view them.

Declination: The angular distance of a celestial body to the plane of the celestial equator.

Ecliptic: The path that the Sun follows over the course of a year as observed from Earth. This path is also the orbit in which the Earth revolves around the Sun.

Ephemeris: A chart listing the coordinates and brightnesses of the stars and planets in the sky.

Equinox: The intersection points of the plane of the celestial equator and the plane of the ecliptic. Night and day are of equal lenght on equinoxes.

Eyepiece (Ocular): The lens directly in front of the eye in systems such as telescopes and microscopes.

Focal Length: The distance from the lens or the primary mirror to the focal point of the telescope, in millimeters. Telescopes with greater focal lengths are more powerful and able to obtain larger images.

Galaxy: Gigantic star systems made up of millions of stars that are linked by gravitation, star clusters, nebulae, gas clouds, and interstellar gas and dust. Galaxies are classified into three types based on their shapes: elliptical, spiral, and irregular.

Galilean Moons: The name given to the 4 largest satellites of Jupiter. They were named "Galilean Moons" because Galileo was the first person to observe them through a telescope.

Greatest Elongation: The maximum angle between the Sun and Mercury or Venus.

Greenwich Observatory: An observatory founded in East London in 1675.

Heliacal Rising: The heliacal rising of a star or planet is when the star or planet becomes visible after passing from behind the Sun. Stars pass from behind the Sun and become visible on the eastern horizon on roughly the same date every year.

Horizon: The boundary at which the sky appears to merge with the Earth or the sea. It is also called the "apparent horizon."

Inferior Conjunction: When an inferior planet comes between the Sun and the Earth, the inferior planet is said to be in inferior conjunction.

Inferior Planet: The name given to the planets located between the Earth and the Sun. Venus and Mercury are inferior planets.

Light-year: The distance that light travels in one year (approximately 10 trillion kilometers).

Lunar Eclipse: A lunar eclipse occurs when the Earth comes between the Sun and the Moon. If the Earth's umbra falls on the Moon, a total lunar eclipse occurs. If the Earth's penumbra falls on the Moon, a partial lunar eclipse occurs.

Lunar Nodes: The points where the orbit of the Moon intersects the ecliptic.

Lunar Phases: Different portions of the Moon are illuminated on a periodic basis as a result of the Moon's revolution around the Earth. These recurring events are known as lunar phases or phases of the Moon.

Meridian: The imaginary line passing through a celestial pole and the zenith. In Latin, *meri* means "mid" and *diem* means "day," therefore, the word *meridian* means "midday."

Meteor Showers: When rock fragments of various sizes enter our atmosphere at great speeds, friction breaks them up into dust, causing traces to form behind them. These occurrences are colloquially known as meteor showers. Meteor showers can produce up to 100 meteors per hour at certain periods of the year. (Millions of meteors are thought to enter the atmosphere every day.) Meteor showers typically occur at heights of 80–110 kilometers in the Earth's atmosphere.

Meteorite: The broad term for small celestial bodies that move in their own orbits within the solar system. Meteorites can range in size from small dust particles to large rock fragments up to a meter in diameter.

The gravitational forces of the planets or the meteorites themselves may cause them to deviate from their orbits and reach our Earth.

Nadir: The lowest point in the apparent orbit of a celestial body. This point is always situated right across from the zenith and is perpendicular to the celestial horizon.

Nebula: Largescale clouds made up of gases such as hydrogen and helium that appear as hazy smudges of light in the sky. If a nebula is near a bright star, the light energy emitted by the star makes the nebula easier to observe. Orion (the hunter), the Magellanic Clouds, Cat's Eye, Horse-head, and Eagle are some of the best-known nebulae.

Opposition: If a superior planet is right across from the Sun as seen from Earth, it is in opposition. Planets are best observed when they are in opposition due to the amount of time they spend in the sky and their brightness.

Orbital Inclination: The inclination of a celestial body's orbit with respect to the plane of the ecliptic.

Perigee: The point where a celestial body is closest to the Earth.

Perihelion: The point where a celestial body is closest to the Sun.

Planet: Celestial bodies that orbit a star and whose masses are too small to produce energy. They get this name because they are mobile.

Pole Star: The bright star near the celestial north pole.

Precession: The shift in the direction of the Earth's axis caused by the Sun's gravitational pull. This is similar to the wobbling of a spinning top. One precessional cycle of the Earth is 26.000 years long, and the pole stars change during this process.

Prime Meridian: The meridian that passes through the Greenwich Observatory.

Satellite: An object that is pulled into orbit around a celestial body by the celestial body's magnetic field.

Sidereal Day: The interval of time between two consecutive meridian transits of a celestial body.

Sidereal Time: The hour angle of the point of Aries. Sidereal time begins when the hour angle of the point of Aries is 0°—i.e., when the point of Aries is at its upper culmination. If the hour angle of any observation location and the right ascension of any star are known, the local sidereal time for that location can be calculated. Thus, sidereal time depends on the local longitude of the observer.

Solar Eclipse: The celestial event that occurs during the new moon phase, when the Moon comes between the Sun and the Earth.

Solstice: The points along the ecliptic where the Sun reaches its maximum distance from the celestial equator.

Space Probe: Robotic spacecrafts tasked with escaping Earth's gravitational field and traveling to the Moon, other celestial bodies, or the interplanetary or intergalactic space to collect data, as well as the tasks performed by these spacecrafts.

Star: Celestial bodies with very large masses that release energy as a result of nuclear reactions in their cores.

Summer Solstice: The first day of summer (approximately June 22). On this date, the Sun rises and sets at its nort-hernmost points.

Sun Dial (Gnomon): A timekeeping tool that uses the shadow of a stick erected perpendicular to the ground to measure the time and is used to determine the motions of the Sun.

Superior Conjunction: When the Sun comes between the Earth and an inferior planet, the inferior planet is said to be in superior conjunction.

Supernova: A phase in the evolution of stars. The violent explosion of a star that results in its components being launched into space is called a supernova.

The Milky Way: One of the billions of galaxies in the observable universe. We are located in the Milky Way.
Usturlab

Vernal Equinox: The first day of spring (approximately March 21). On this date, the weather begins to get warmer in the northern hemisphere.

White Dwarf: A small star with a diameter of several thousand kilometers and a surface temperature of roughly 100,000°. This is the final stage in the evolution of Sun-like stars.

Winter Solstice: The first day of winter (approximately December 22). On this date, days begin to get shorter in the northern hemisphere.

Zenith: The point at which a virtual line extending straight overhead of an observer intersects the sky.

Zodiac: The belt that covers 8°-9° degrees on either side of the ecliptic. It contains the Sun, the Moon, and the planets.

Name: Birth Place:

Date of Birth: Latitude:

The Hour of Birth Longitude:

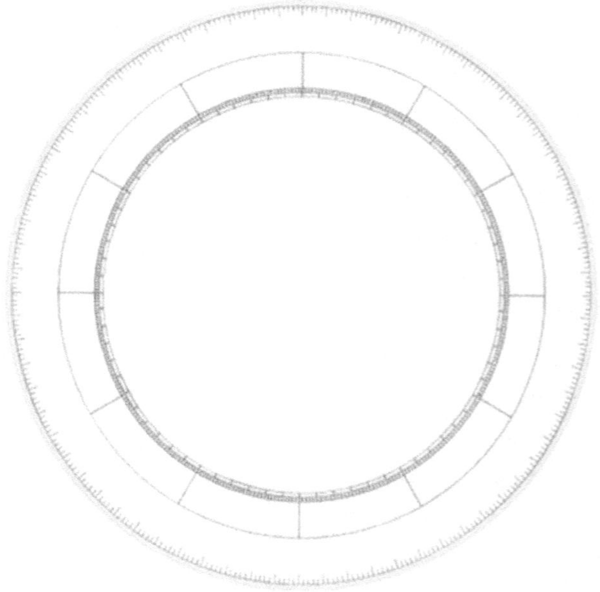

Name: Birth Place:

Date of Birth: Latitude:

The Hour of Birth Longitude:

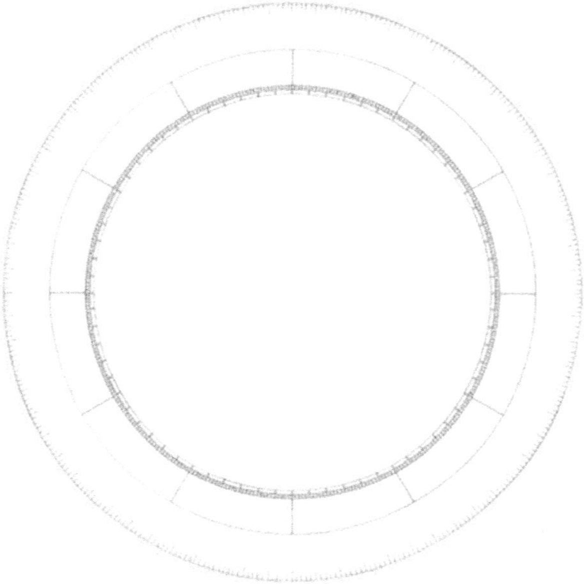

References

Gökyüzü Tarihi, Tarih Boyunca Araçlar ve İmgeler, Boyut Yayınevi, 2009

Astronomi Dersler Cilt 1, Doç. Dr. Abdullah Kızılırmak, Küresel *Astronomi ve Gök Mekaniği,* Ege Üniversitesi Fen Fakültesi Kitaplar Serisi No:11, İzmir, 1964

Astronomi Hazineleri, Muammer Dizer, Boğaziçi Üniversitesi Yayınları No:404,

Gökbilim Dersleri Cilt 1, Küresel Gökbilim, 2. Baskı. Prof. Dr. Abdullah Kızılırmak, Ege Üniversitesi Fen Fakültesi Kitaplar Serisi No:72

Türk-İslam Astronomi Bilginleri ve Gökyüzü Bilgileri, Lütfi Göker, Milli Eğitim Bakanlığı Yayınları, Bilim ve Kültür Eserleri Dizisi: 742, 1995, İstanbul

Natal Charting, John Filbey, An Aquarian Astrology Handbook, 1981

Lecture Notes for Introduction to Astronomy, Denver Museum of Nature&Science 2005, Ka Chun Yu

The Mathematics of Astrology, Does House Division Make Sense?, Kevin Heng Ser Guan

Department of Physics National University of Singapore, 2000-2001

Astronomide Kullanılan Teleskoplar ve Bunların Çalışma Prensipleri, İbrahim Gündemir, Ege Üniversitesi Fen Bilimleri Enstitüsü, Astronomi ve Uzay Bilimleri Anadalı, 1990

Takımyıldızlar ve Yıldız Haritaları, Prof. Dr. Serdar Evren, Ege Üniversitesi Astronomi Topluluğu Yardımcı Yayınlar Dizisi, No:13 1997

Popüler Astronomi, Facebook, Prof. Dr. Serdar Evren

Osmanlı Astronomisine Genel Bir Bakış, Dr. Yavuz Unat

İlk Ve Orta Öğretimde Astronomi Uygulamaları, Ankara Üniversitesi Fen Bilimleri Enstitüsü Yüksek Lisans Tezi, Sevda Sakallı, 2008

Güneş Ötesi Gezegen Sistemleri, Çanakkale Onsekiz Mart Üniversitesi Fen Bilimleri Enstitüsü Uzay Bilimleri ve Teknolojileri Anabilim Dalı, Yüksek Lisans Tezi, Arif Solmaz, 2010

Ay ve Güneş Tutulmaları Ders Notu, Selim O. Selam, Ankara Üniversitesi Fen Fakültesi Astronomi ve Uzay Bilimleri Bölümü, 2007

Gökyüzü Gözlemciliği, Tübitak Popüler Bilim Yayınları, Alp Akoğlu

Gezegenlerin Hareketleri, H. Turgay Kaptanoğlu, Odtü Matematik Bölümü

Fen Eğitiminde Astronomi Uygulamalar, Ayşe Zurnacı, 2015, Ege Üniversitesi Fen Bilimleri Enstitüsü Yüksek Lisans Tezi

Yıldızlardan Yıldızlara Astronomi Semineri, Yrd. Doç. Dr. Emre Işık, İstanbul Kültür Üniversitesi Fen- Edebiyat Fizik Bölümü, 2012

Prof.Dr. Abdullah Kızılırmak, Küresel Astronomi Ders Kitabı

Prof. Dr. Berahitdin Albayrak, Genel Astronomi Ders Notları

Tübitak Bilim ve Teknik Dergisi, Yıldız Takımı, Sayı 9, Haziran 2009, Neden Astronomi,

Astronomi ve Uzay Bilimleri, Milli Eğitim Bakanlığı Devlet Kitapları, 2012

Bilim Tarihi, Colin A. Ronan, Tübitak Akademik Dizi, 2003

Güneş Sisteminin Temel Fiziği, Ankara Üniversitesi Fen Fakültesi Yayınları, V.M. Blanco ve S.W. McCuskey, Ankara, 2018

Eski Çağlardan Bu Yana Ölçümü ve Takvim Prof. Dr. Neşet Çağatay

National Geography Planet Belgeselleri

Gökyüzü, Türk Astronomi Derneği Aylık Bülteni, Temmuz 2015 Sayı 59

Gökyüzü, Türk Astronomi Derneği Aylık Bülteni, Mart 2014, Sayı 51

Evren Atlası, Mark A. Garlick, NTV Yayınları

Yıldızların zamanı, Alan Lightman

Mark A. Garlick'in The Expanding Universe

Tarih Boyunca Geliştirilmiş Evren Modelleri, Bilim Tarihinden, Prof. Dr. Gazi Topdemir

Güneş Merkezli Evren Anlayışı: Kopernik, Kepler ve Gallilei Neyi Değiştirdi? Seda Özsoy

Gezegenler Kılavuzu, Patrick Moore, Tübitak Popüler Bilim Kitapları, 1999

Paleolitik Çağ Kadınları, Neyir Kolankaya Bostancı

Whole Sign Houses, The Oldest House System, Robert Hand, Arhat Publication, 2000

http://asart.science.ankara.edu.tr/gunes_sistemi/ay.php

Retrograde Planets, Erin Sullivan

Burçlar, Devrim Dölen& Devrim Yılmazer, Zodyak Astroloji Yayınevi

Evler, Devrim Dölen& Devrim Yılmazer, Zodyak Astroloji Yayınevi

Takvim: Zamanın Haritası, Prof. Dr. Çiğdem Dürüşken, 2000, Cogito, sayı:22 Yapı Kredi Yayınları

References of Images

Image 1. *https://sarahfuhro.files.wordpress.com/2014/06/lunar-calendar1.jpg*

Image 2. *https://cdn3.bigcommerce.com/s-290/images/476/Venus-Laussel-Dordogne-France-22__27946.1441480015.500.750.jpg?c=2*

Image 3. *https://www.bilimma.com/gobeklitepe-din-arkeolojisi-nedir/*

Image 4. *https://www.bilimma.com/gobeklitepe-din-arkeolojisi-nedir/*

Image 5. *https://encrypted-tbn0.gstatic.com/images*

Image 6. *https://encrypted-tbn0.gstatic.com/images*

Image 7. *https://encrypted-tbn0.gstatic.com/images?q=tbn:ANd9G-cRkIM418nliJ7wPYXawmQj8Uo894dMq54jv3yTCHNWWsoIexrd8*

Image 8. *http://media-2.web.britannica.com/eb-media//52/752-050-CA91D3CB.jpg*

Image 9. *http://www.matematiksel.org/wp-content/uploads/2016/02/s-a0da725b61fc2ff7bf08ba1751baa8e07f2864d8.jpg*

Image 10. *http://ektalks.blogspot.com.tr/2016/02/how-old-is-milky-way-and-how-do-we-know.html*

Image 11. *https://encrypted-tbn0.gstatic.com/images?q=tbn:ANd9GcT-4bV_pwlY8M65tVWJQqJvKuEoCFBQwxOCE9cFOhafQPQHz2PE_*

Image 12. *http://www.astronoo.com/images/planetes/axial-tilt-planets.jpg*

Image 13. *https://encrypted-tbn0.gstatic.com/images?q=tbn:ANd9GcT-070BvX3ytQG08Y0Wuo54Y9u_8ZQ4IKEf6PRu527gSjKaXRLBl*

Image 14. *https://1.bp.blogspot.com/-Nwx5_OQwfC0/Teaz9d0T8vI/AAAAAAAAAYk/OIecn4bjm6s/s640/zodiac_clock.gif*

Image 15. *https://encrypted-tbn0.gstatic.com/images?q=tbn:ANd9GcRbC-mc0QNPyh4zNOo7aZcKEne1Qd79LE0FzjQNY2Lef5FYQYJNYsg*

Image 16. *http://media1.britannica.com/eb-media/68/91868-034-8485B5D3.gif*

Image 17. *https://upload.wikimedia.org/wikipedia/commons/8/83/Perihelium-aphelium.GIF*

Image 18. *From Solar Fire programme*

Image 19. *From Solar Fire programme*

Image 20. *https://encrypted-tbn0.gstatic.com/images?q=tbn:ANd9GcQZ-jtzFDnHauFb7EFimdM0vPhz_HSfnEV1z2OaWTHi4xOA-gw-z*

Image 21. *http://www.ccg.org/weblibs/study-papers/Image1065.gif*

Image 22. *http://www.douglasvincent.com/photography/star-trails/images/star-trails-finding-polaris.jpg*

Image 23. *http://atmosinfo.com/wp-content/uploads/2016/12/Greek-Alphabet-Letters-and-Names.png*

Image 24. *http://www.eyb.com.tr/gok-atlas-takvimi*

Image 25. *http://www.telescopescene.com/refracting-telescope-guides/*

Image 26. *https://lco.global/spacebook/reflecting-telescopes/*

Image 27. *https://upload.wikimedia.org/wikipedia/commons/thumb/5/5a/Schmidt-Cassegrain-Teleskop.svg/1280px-Schmidt-Cassegrain-Teleskop.svg.png*

Image 28. *https://encrypted-tbn0.gstatic.com/images?q=tbn:ANd9GcSkq-2tA_Bw8Fmm_6VxLKay6aYxK7LcYuJg7ZtjecbfGxPgIowFS*

Image 29. *https://www.astro.com/swisseph/swepha_e.htm*

Made in United States
Cleveland, OH
05 November 2026

47772394R00112